Specific Heats of Various Materials[§]

Substance	Specific Heat ($J\ kg^{-1}\ K^{-1}$)
Elemental Solids:	
Lead	128
Gold	129
Silver	234
Copper	387
Iron	448
Aluminum	900
Other Solids:	
Granite	760
Glass	837
Marble	860
Wood	1700
Ice (−10°C)	2200
Liquids:	
Mercury	139
Ethyl Alcohol	2430
Water	4186
Air	≈ 740

[§]Measured at 300 K unless otherwise noted. Based on Serway, *Physics,* 3d ed., p. 530.

Latent Heats of Various Substances[¶]

At Standard Pressure

Substance	Melting Point (K)	Latent Heat of Fusion (kJ/kg)	Boiling Point (K)	Latent Heat of Vaporization (kJ/kg)
Helium	(does not solidify at 1 atm)		4.2	21
Hydrogen	14.0	58.6	20.3	452
Oxygen	54.8	13.8	90.2	213
Nitrogen	63.2	25.5	77.4	201
Mercury	234	11.3	630	296
Water	273	333	373	2256
Lead	601	24.7	2013	853
Aluminum	933	105	2720	11,400
Copper	1356	205	2840	4730

[¶]Adapted from Halliday, Resnick, and Krane, *Physics,* 4th ed., Wiley, New York, p. 550.

Six Ideas That Shaped Physics

Unit T: Some Processes Are Irreversible

Physics

Second Edition

Thomas A. Moore

McGraw Hill

Boston   Burr Ridge, IL   Dubuque, IA   Madison, WI   New York   San Francisco   St. Louis
Bangkok   Bogotá   Caracas   Kuala Lumpur   Lisbon   London   Madrid   Mexico City
Milan   Montreal   New Delhi   Santiago   Seoul   Singapore   Sydney   Taipei   Toronto

McGraw-Hill Higher Education

*A Division of The **McGraw-Hill** Companies*

SIX IDEAS THAT SHAPED PHYSICS, UNIT T: SOME PROCESSES ARE IRREVERSIBLE
SECOND EDITION

4 5 6 7 8 9 0 QPD/QPD 0 9 8 7

ISBN   978-0-07-239715-4
MHID 0-07-239715-2

Publisher: *Kent A. Peterson*
Sponsoring editor: *Daryl Bruflodt*
Developmental editor: *Spencer J. Cotkin, Ph.D*
Marketing manager: *Debra B. Hash*
Senior project manager: *Susan J. Brusch*
Senior production supervisor: *Sandy Ludovissy*
Coordinator of freelance design: *David W. Hash*
Cover/interior designer: *Rokusek Design*
Cover image: *©Nova Development*
Senior photo research coordinator: *Lori Hancock*
Photo research: *Chris Hammond/PhotoFind LLC*
Supplement producer: *Brenda A. Ernzen*
Media project manager: *Sandra M. Schnee*
Media technology associate producer: *Judi David*
Compositor: *Interactive Composition Corporation*
Typeface: *10/12 Palatino*
Printer: *Quebecor World Dubuque, IA*

Credit List:
Chapter 1 T1.6: © Frozen Images/Image Works. **Chapter 2** T2.5: © Corbis/Vol. 1; T2.6: NASA. **Chapter 3** T3.10: © National Center For Atmospheric Research/University Corporation for Atmospheric Research/National Science Foundation; Pg. 56(right): © Vivian Peevers/Peter Arnold. **Chapter 5** T5.7: © McGraw-Hill Companies, Inc., photo by C. P. Hammond; T5.8: © Peter Vandermark/Stock Boston; Pg. 91: April 1997 cover, © 1997 by Scientific American, Inc. All rights reserved. **Chapter 6** Page 110: Courtesy of NRAO/AUI. **Chapter 7** Page 132: © Karen Su/Stock Boston. **Chapter 8** T8.3,T8.5: © Richard Megna/Fundamental Photographs. **Chapter 9** T9.4: Courtesy Niagara Mohawk; T9.5 © by John Wiley & Sons. Used by permission of John Wiley & Sons, Inc.; T9.6: National Oceanic and Atmospheric Administration/Department of Commerce; T9.8: Courtesy of Lennox Industries Inc.; T9.15: Courtesy of Leybold Didactic, GMBH.

Library of Congress Cataloging-in-Publication Data

Moore, Thomas A. (Thomas Andrew)
  Six ideas that shaped physics. Unit T, Some processes are irreversible / Thomas A. Moore. — 2nd ed.
     p. cm.
  Includes index.
  ISBN 0-07-239715-2 (acid-free paper)
  1. Irreversible processes. I. Title.

QC318.17 .M66   2003
536'.7—dc21   2001056201
                                    CIP

www.mhhe.com

Dedication

For Allison,
whose warmth is legendary.

Table of Contents for
Six Ideas That Shaped Physics

Contents: Unit T
Some Processes Are Irreversible

Chapter T7

Some Mysteries Resolved   112

Chapter T9

Heat Engines   154

Chapter T8

Calculating Entropy Changes   134

About the Author

Thomas A. Moore graduated from Carleton College (*magna cum laude* with Distinction in Physics) in 1976. He won a Danforth Fellowship that year that supported his graduate education at Yale University, where he earned a Ph.D. in 1981. He taught at Carleton College and Luther College before taking his current position at Pomona College in 1987, where he won a Wig Award for Distinguished Teaching in 1990. He served as an active member of the steering committee for the national Introductory University Physics Project (IUPP) from 1987 through 1995. This textbook grew out of a model curriculum that he developed for that project in 1989, which was one of only four selected for further development and testing by IUPP.

He has published a number of articles about astrophysical sources of gravitational waves, detection of gravitational waves, and new approaches to teaching physics, as well as a book on special relativity entitled "A Traveler's Guide to Spacetime" (McGraw-Hill, 1995). He has also served as a reviewer and an associate editor for *American Journal of Physics*. He currently lives in Claremont, California, with his wife Joyce and two college-aged daughters, Brittnay and Allison. When he is not teaching, doing research in relativistic astrophysics, or writing, he enjoys reading, hiking, scuba diving, teaching adult church-school classes on the Hebrew Bible, calling contradances, and playing traditional Irish fiddle music.

Preface

Introduction

This volume is one of six that together comprise the text materials for *Six Ideas That Shaped Physics*, a fundamentally new approach to the two- or three-semester calculus-based introductory physics course. *Six Ideas That Shaped Physics* was created in response to a call for innovative curricula offered by the Introductory University Physics Project (IUPP), which subsequently supported its early development. In its present form, the course represents the culmination of more than a decade of development, testing, and evaluation at a number of colleges and universities nationwide.

Opening comments about *Six Ideas That Shaped Physics*

This course is based on the premise that innovative approaches to the presentation of topics and to classroom activities can help students learn more effectively. I have completely rethought from the ground up the presentation of every topic, taking advantage of research into physics education wherever possible, and I have done nothing just because "that is the way it has always been done." Recognizing that physics education research has consistently underlined the importance of active learning, I have also provided tools supporting multiple opportunities for active learning both inside and outside the classroom. This text also strongly emphasizes the process of building and critiquing physical models and using them in realistic settings. Finally, I have sought to emphasize contemporary physics and present even classical topics from a thoroughly contemporary perspective.

I have not sought to "dumb down" the course to make it more accessible. Rather, my goal has been to help students become *smarter*. I intentionally set higher-than-usual standards for sophistication in physical thinking, and I have then used a range of innovative approaches and classroom structures to help even average students reach this standard. I do not believe that the mathematical level required by these books is significantly different from that of a standard university physics text, but I do ask students to step beyond rote thinking patterns to develop flexible, powerful conceptual reasoning and model-building skills. My experience and that of other users are that normal students in a wide range of institutional settings can, with appropriate support and practice, meet these standards.

The six volumes that comprise the complete *Six Ideas* course are:

The six volumes of the *Six Ideas* text

Unit C (**C**onservation Laws):	Conservation Laws Constrain Interactions
Unit N (**N**ewtonian Mechanics):	The Laws of Physics are Universal
Unit R (**R**elativity):	The Laws of Physics are Frame-Independent
Unit E (**E**lectricity and Magnetism):	Electricity and Magnetism are Unified
Unit Q (**Q**uantum Physics):	Matter Behaves Like Waves
Unit T (**T**hermal Physics):	Some Processes are Irreversible

I have listed these units in the order that I recommend that they be taught, though other orderings are possible. At Pomona, we teach the first three units during the first semester and the last three during the second semester of a year-long course, but one can easily teach the six units in three quarters or even over three semesters if one wants a lower pace.

The chapters of all of these texts have been designed to correspond to the *maximum* amount that one might realistically discuss in a single 50-minute class session. While one might design a syllabus that covers chapters at a slower rate, one should *not* try to discuss more than one chapter in 50-minute class. A list of the chapters in all six volumes of the course appears on pp. vii–viii.

For more information than I can include in this short preface about the goals of the *Six Ideas* course, its organizational structure (and the rationale behind that structure), the evidence for its success, and information about how to cut and/or rearrange material, as well as many other resources for both teachers and students, please visit the *Six Ideas* web site (see the next section).

Important Resources

Instructions about how to use this text

The *Six Ideas* web site

I have summarized important information about how to read and use this text in a section entitled *Introduction for Students* that starts on page xiv. Please look this over, particularly if you have not seen previous volumes of this text.

The *Six Ideas* web site contains a wealth of up-to-date information about the course that I think both instructors and students will find very useful. The URL is

www.physics.pomona.edu/sixideas/

Essential computer programs

One of the most important resources available at this site is a number of computer applets that illustrate important concepts and aid in difficult calculations. In critical places, this unit draws on several of these programs, and past experience indicates that students learn the ideas much more effectively when these programs are used both in the classroom and for homework. These applets are freeware and are available for both the Macintosh (Classic) and Windows operating systems.

Notes Specifically About Unit T

How this unit depends on the other units

In the suggested *Six Ideas* unit sequence, this unit is the last of the six. This is done partly so that the course ends with something pretty easy and yet interesting and partly so that students have seen the quantum physics in unit Q, particularly the material regarding the simple harmonic oscillator and the particle in a box.

However, this is not crucial. Students seem to readily accept that quantum systems have energy levels, and that is the only idea that is really important. Some aspects of the particle-in-a-box model are used at the beginning of chapters T7 and T8, but all that is *really* necessary is for students to know and accept that when a particle is confined to a box, its velocity components are quantized in units of $h/2mL$. In the class session before they read chapter T7, you might state this as given and have them skip section T7.2. Alternatively, you can simply spend some class time teaching the particle-in-a-box model if students have not seen it before (to be extra safe, you might budget an extra class day to do this).

The only real *requirement* is that this unit follows unit C. This unit does not really draw on anything crucial in any of the other units.

How to make cuts if necessary

As in all the *Six Ideas* texts, the chapters in this unit are designed to correspond to one 50-minute class session. This sets a pretty stiff pace, though, and your students may benefit from a slower pace (which you can arrange by scheduling one class session every one or two weeks for consolidation and review). However, do not attempt to cover more than one chapter per 50-minute class session.

If you need to cut chapters, cut chapter T7 first: this is not needed for any other chapter (it does resolve a few of the mysteries raised in chapter T2, though). If you are willing to omit essentially *all* applications, you could also cut chapters T8 and T9 (there is not much point in discussing chapter T8 without chapter T9). Chapters T1 through T6, however, form an essentially indivisible whole.

Experience has shown that the ideas discussed in chapter T5 make *much* more sense to students (and also have a greater impact) when some class and homework time is spent using the computer program StatMech. (This is the only unit in the course that really almost *requires* the help of a computer program to function well.) Please download this program from the *Six Ideas* website; it is freeware, so you may distribute it freely.

Use the computer program StatMech

We have found that this unit does provide a nice note on which to end the course. The concepts are not that difficult, but they are important and really have a profound connection with daily life. Moreover, covering a relatively straight-forward unit near the end of the term (when life is usually very stressful) helps students keep up and maintain a positive attitude about the course.

Appreciation

A project of this magnitude cannot be accomplished alone. First, I would like to thank the others who served on the IUPP development team for this project: Edwin Taylor, Dan Schroeder, Randy Knight, John Mallinckrodt, Alma Zook, Bob Hilborn, and Don Holcomb. I would like to thank John Rigden and other members of the IUPP steering committee for their support of the project in its early stages, which came ultimately from a National Science Foundation grant and the special efforts of Duncan McBride. Users of the texts—especially Bill Titus, Richard Noer, Woods Halley, Paul Ellis, Doreen Weinberger, Nalini Easwar, Brian Watson, Jon Eggert, Catherine Mader, Paul De Young, Alma Zook, Dan Schroeder, David Tanenbaum, Alfred Kwok, and Dave Dobson—have offered invaluable feedback and encouragement. I also would like to thank Alan Macdonald, Roseanne Di Stefano, Ruth Chabay, Bruce Sherwood, and Tony French for ideas, support, and useful suggestions. Thanks also to Robs Muir for helping with several of the indexes. My editors Jim Smith, Denise Schanck, Jack Shira, Karen Allanson, Lloyd Black, J. P. Lenney, and Daryl Bruflodt as well as Spencer Cotkin, Donata Dettbarn, David Dietz, Larry Goldberg, Sheila Frank, Jonathan Alpert, Zanae Roderigo, Mary Haas, Janice Hancock, Lisa Gottschalk, Debra Hash, David Hash, Patricia Scott, Chris Hammond, Rick Hecker, Brittney Corrigan-McElroy, and Susan Brusch have all worked very hard to make this text happen, and I deeply appreciate their efforts. I would like to thank reviewers Brenda Weiss (University of California, Davis), Xiang Ning Song (Richland College), Victor DeCarlo (De Pauw University), Derrick Hylton (Spelman College), Edwin Carlson, David Dobson, Irene Nunes, Miles Dressler, O. Romulo Ochoa, Qichang Su, Brian Watson, and Laurent Hodges for taking the time to do a careful reading of various units and offering valuable suggestions. Thanks to Connie Wilson, Hilda Dinolfo, Connie Inman, and special student assistants Michael Wanke, Paul Feng, Mara Harrell, Jennifer Lauer, Tony Galuhn, Eric Pan, and all the Physics 51 mentors for supporting (in various ways) the development and teaching of this course at Pomona College. Thanks also to my Physics 51 students, and especially Win Yin, Peter Leth, Eddie Abarca, Boyer Naito, Arvin Tseng, Rebecca Washenfelder, Mary Donovan, Austin Ferris, Laura Siegfried, and Miriam Krause, who have offered many suggestions and have together found many hundreds of typos and other errors. Eric Daub was indispensable in helping me put this edition together. Finally, very special thanks to my wife Joyce and to my daughters Brittany and Allison, who contributed with their support and patience during this long and demanding project. Heartfelt thanks to all!

Thanks!

Thomas A. Moore
Claremont, California

Introduction for Students
How to Read and Use This Text Effectively

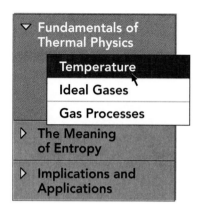

Introduction

Welcome to *Six Ideas That Shaped Physics!* This text has been designed using insights from recent research into physics learning to help you learn physics as effectively as possible. It thus has many features that may be different from those in science texts you have probably encountered. This section discusses these features and how to use them effectively.

Why Is This Text Different?

Research consistently shows that people learn physics most effectively if they participate in *activities* that help them *practice* applying physical reasoning in realistic situations. This is so because physics is not a collection of facts to absorb, but rather is a set of *thinking skills* requiring practice to master. You cannot learn such skills by going to factual lectures any more than you can learn to play the piano by going to concerts!

This text is designed, therefore, to support *active learning* both inside and outside the classroom by providing (1) resources for various kinds of learning activities, (2) features that encourage active reading, and (3) features that make it easier for the text (as opposed to lectures) to serve as the primary source of information, so that more class time is available for active learning.

The Text as Primary Source

Features that help the text serve as the primary source of information

To serve the last goal, I have adopted a conversational style that I hope will be easy to read, and I tried to be concise without being so terse that you need a lecture to fill in the gaps. There are also many text features designed to help you keep track of the big picture. The unit's **central idea** is summarized on the front cover where you can see it daily. Each chapter is designed to correspond to one 50-minute class session, so that each session is a logically complete unit. The two-page **chapter overview** beginning each chapter provides a compact summary of that chapter's contents to consider before you are submerged by the details (it also provides a useful summary when you review for exams). An accompanying **chapter location diagram** uses a computer menu metaphor to display how the current chapter fits into the unit (see the example at the upper left). Major unit subdivisions appear as gray boxes, with the current subdivision highlighted in color. Chapters in the current subdivision appear in a submenu with the current chapter highlighted in black and indicated by an arrow.

All technical terms are highlighted using a **bold** type when they first appear, and a **glossary** at the end of the text summarizes their definitions. Please also note the tables of useful information, including definitions of common symbols, that appear inside the front cover.

A physics *formula* is both a mathematical equation and a *context* that gives the equation meaning. Every important formula in this text appears in a **formula box.** Each contains the equation, a **purpose** (describing the formula's meaning and utility), a definition of the **symbols** used in the equation, a description of any **limitations** on the formula's applicability, and possibly some other useful **notes.** Treat everything in such a box as an *indivisible unit* to be remembered and used together.

Active Reading

Like passively listening to a lecture, passively scanning a text does not really help you learn. *Active* reading is a crucial study skill for effectively learning from this text (and other types of technical literature as well). An active reader stops frequently to pose internal questions such as these: *Does this make sense? Is this consistent with my experience? Am I following the logic here? Do I see how I might use this idea in realistic situations?* This text provides two important tools to make this easier.

Use the **wide margins** to (1) record *questions* that occur to you as you read (so that you can remember to get them answered), (2) record *answers* when you receive them, (3) flag important passages, (4) fill in missing mathematics steps, and (5) record insights. Doing these things helps keep you actively engaged as you read, and your marginal comments are also generally helpful as you review. Note that I have provided some marginal notes in the form of *sidebars* that summarize the points of crucial paragraphs and help you find things quickly.

The **in-text exercises** help you develop the habits of (1) filling in missing mathematics steps and (2) posing questions that help you *practice* using the chapter's ideas. Also, although this text has many examples of worked problems similar to homework or exam problems, *some* of these appear in the form of in-text exercises (as you are more likely to *learn* from an example if you work on it some yourself instead of just scanning someone else's solution). Answers to *all* exercises appear at the end of each chapter, so you can get immediate feedback on how you are doing. Doing at least some of the exercises as you read is probably the *single most important thing you can do* to become an active reader.

Active reading does take effort. *Scanning* the 5200 words of a typical chapter might take 45 minutes, but active reading could take twice as long. I personally tend to "blow a fuse" in my head after about 20 minutes of active reading, so I take short breaks to do something else to keep alert. Pausing to fill in missing mathematics also helps me to stay focused longer.

Class Activities and Homework

The problems at the end of each chapter are organized into categories that reflect somewhat different active-learning purposes. **Two-minute problems** are short, concept-oriented, multiple-choice problems that are primarily meant to be used *in* class as a way of practicing the ideas and/or exposing conceptual problems for further discussion. (The letters on the back cover make it possible to display responses to your instructor.) The other types of problems are primarily meant for use as homework *outside* class. **Basic** problems are simple drill-type problems that help you practice in straightforward applications of a single formula or technique. **Synthetic** problems are more challenging and realistic questions that require you to bring together multiple formulas

---

What it means to be an active *reader*

Tools to help you become an active reader

The single most important thing you can do

End-of-chapter problems support active learning

and/or techniques (maybe from different chapters) and to think carefully about physical principles. These problems define the level of sophistication that you should strive to achieve. **Rich-context** problems are yet more challenging problems that are often written in a narrative format and ask you to answer a practical, real-life question rather than explicitly asking for a numerical result. Like situations you will encounter in real life, many provide too little information and/or too much information, requiring you to make estimates and/or discard irrelevant data (this is true of some *synthetic* problems as well). Rich-context problems are generally too difficult for most students to solve alone; they are designed for *group* problem-solving sessions. **Advanced** problems are very sophisticated problems that provide supplemental discussion of subtle or advanced issues related to the material discussed in the chapter. These problems are for instructors and truly exceptional students.

Read the Text *Before* Class!

Class time works best if you are prepared

You will be able to participate in the kinds of activities that promote real learning *only* if you come to each class having already read and thought about the assigned chapter. This is likely to be *much* more important in a class using this text than in science courses you may have taken before! Class time can also (*if* you are prepared) provide a great opportunity to get your *particular* questions about the material answered.

T1

Temperature

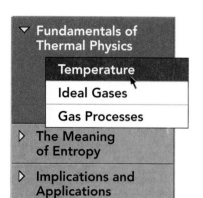

Chapter Overview

Section T1.1:   Introduction to the Unit

Thermodynamics is the study of how a macroscopic object's temperature, thermal energy, entropy, and similar *macroscopic* properties are affected by its interactions with other objects and its general environment. **Statistical mechanics** is the theory that explains the thermodynamic behavior of complex systems in terms of the *statistical* behavior of molecules obeying simple laws of mechanics. We will explore the foundations of statistical mechanics in this unit.

Section T1.2:   Irreversible Processes

The simple interactions between molecules at the microscopic level are **reversible processes,** in the sense that a reversed movie of the interaction also depicts a physically possible interaction. Many processes involving macroscopic objects, however, are **irreversible processes** in that sense. One of the challenges faced by statistical mechanics is to explain how irreversible macroscopic behavior arises out of reversible microscopic behavior.

This unit is divided into three subunits of three chapters each. The first subunit provides important background by defining core thermodynamic concepts and exploring a simple newtonian model of a gas. The second uses a simple quantum model of a solid to present Boltzmann's solution to the problem of irreversibility. The third explores practical consequences of these ideas.

Section T1.3:   The Paradigmatic Thermal Process

When a hot object is placed in contact with a cold object, heat flows irreversibly from the former to the latter until they come into **thermal equilibrium** (their thermal properties no longer change with time). Our primary goal in this unit will be to explain this **paradigmatic thermal process.** Understanding this problem will force us to resolve a number of critical issues regarding the nature of temperature, equilibrium, and irreversibility.

Section T1.4:   Temperature and Equilibrium

A **thermoscope** is an object having a measurable property whose numerical value is uniquely linked (somehow) to the object's temperature. Thermoscope measurements empirically show that objects interacting thermally obey the **zeroth law of thermodynamics:**

> A well-defined quantity called **temperature** exists such that two objects in thermal contact will be in thermal equilibrium *if and only if* both have the same temperature (i.e., thermoscope reading).

This expresses the core physical meaning of temperature. In chapter T6 we will see how we can *define* temperature in terms of more basic concepts in a way automatically consistent with this law.

Section T1.5:   Thermometers

Historically, however, temperature was first defined quantitatively by picking the **constant-volume gas thermoscope**˙to serve as a standard. An object's absolute temperature T is *defined* to be proportional to the **pressure** P (force per unit area) exerted by the gas on the walls of a constant-volume thermoscope in thermal equilibrium with the object. Note that this definition implies that $T = 0$ when $P = 0$. Since a gas can never exert a negative pressure, this is the lowest possible temperature (**absolute zero**).

To define the constant of proportionality between P and T, we need to specify the numerical temperature of some easily reproduced standard object or situation. It turns out that water vapor, liquid water, and ice can coexist only at a unique temperature, which we call water's **triple point** (TP). We define this unique temperature to be 273.16 K, where K is the symbol for the **kelvin,** the SI (*Système International*) temperature unit. Then an object's absolute temperature is

$$T \equiv 273.16\,\text{K} \left(\lim_{\rho \to 0} \frac{P}{P_{\text{TP}}} \right) \qquad \text{(T1.3)}$$

> **Purpose:** This equation empirically defines an object's temperature T in kelvins.
> **Symbols:** P is the gas pressure in a constant-volume gas thermoscope in equilibrium with the object, and P_{TP} is the pressure of the same thermoscope in equilibrium with water at its triple point.
> **Limitations:** The limit symbol indicates that we should evaluate the ratio in the limit that the gas density in the thermoscope goes to zero. This equation defines temperatures on the Kelvin scale only.
> **Notes:** Taking the zero-density limit is useful because all types of gases behave identically in this limit.

A **thermometer** is a thermoscope calibrated to register the same numerical temperature as a constant-volume gas thermoscope would.

Equation T1.3 defines the Kelvin temperature scale. We can define different temperature scales by redefining the scale's zero point and/or the size of the temperature unit. We can convert a temperature from one scale to another by (1) symbolically expressing the temperature on both scales as a temperature *difference* from some reference point whose temperature you know on both scales, (2) multiplying the difference on the original scale by a unit operator to convert it to a difference on the new scale, and (3) solving for the temperature value on the new scale.

Section T1.6: Temperature and Thermal Energy

The temperature T of most objects increases as we add energy to an object (increasing its thermal energy U). An object's thermal energy is linked to its temperature by

$$dU = mc\,dT \qquad \text{or} \qquad c \equiv \frac{1}{m}\frac{dU}{dT} \qquad \text{(T1.6)}$$

> **Purpose:** This equation describes the empirically observed link between the change dU in an object's internal energy and its corresponding change in temperature dT.
> **Symbols:** m is the object's mass, and c is the object's **specific heat** (*not* the speed of light!).
> **Limitations:** dU and dT must be small enough so that $c \approx$ constant. The value of c is often undefined during a phase change.
> **Notes:** The second version of this equation amounts to a definition of c. Specific "heat" is a horrible misnomer here, because we can add energy in forms other than heat.

This equation is useful partly because c often has a nearly fixed value for a given type of substance in a given phase.

T1.1   Introduction to the Unit

Complex objects have their own distinct physics

In much of this course to date, we have treated objects (from atoms to automobiles to planets) as if they were pointlike particles. Yet every object that we experience in our daily lives is in fact a complicated system comprised of an enormous number of tiny particles. We laid some of the foundations for studying complex systems in unit C, but in this unit we will study the behavior of such systems in much greater depth.

Complex systems have physical properties we would not think of assigning to point particles. We have already discussed in unit C how complex systems have *thermal energy* that must be considered when applying the law of conservation of energy. This thermal energy is related in some (as yet mysterious) way to the object's *temperature*, also something a point particle does not have. In this unit, we see that complex systems also have something called *entropy*, but particles do not.

We will see in each case that such quantities are ultimately *statistical* expressions of the internal structure or "state" of a complex system. For example, the link between temperature and a system's internal state is vividly illustrated by the freezing of water. Clearly something quite dramatic happens to the internal state of a system of water molecules when its temperature is lowered below 0°C.

Physicists used the concepts of *temperature, heat, thermal energy,* and *entropy* to describe the behavior of everyday objects long before they understood such objects to be systems of molecules. The concept of *temperature*, for example, is nearly as old as humanity, since we directly sense when something is hot or cold. We also know from common experience that when we put a cold object on a fire, something flows to it from the fire that makes it hot. Even the idea that this "something" was a form of energy was discovered in the 1830s, many decades before scientists generally accepted the atomic hypothesis of matter.

Thermodynamics and statistical mechanics

The study of how a macroscopic object's temperature, thermal energy, entropy, and similar *macroscopic* characteristics are affected by its interactions with other objects and its general environment is called **thermodynamics,** which became a well-developed science in the middle decades of the 1800s. One of the greatest triumphs of physics in the early decades of the 1900s was the development of a theory called **statistical mechanics** that explains the thermodynamic behavior of systems in terms of the *statistical* behavior of molecules obeying the simple laws of mechanics. We will focus our attention mostly on the principles of statistical mechanics in this unit.

T1.2   Irreversible Processes

Historically, the greatest impediment to the development of statistical mechanics was a particular stark difference between the behavior of macroscopic objects and the behavior of point particles: the behavior of point particles is generally **reversible,** while the behavior of macroscopic objects is often **irreversible.**

Reversible processes

Imagine, for example, making a movie of two perfectly elastic billiard balls colliding on a pool table. The movie will show the balls approaching each other, exchanging some energy and momentum in the collision process, and then receding from each other. If we were to run the movie backward, we would also see the balls approaching each other, exchanging some energy and momentum in the collision process, and then receding from each other. As long as the collision is perfectly elastic (so that the thermal energy of the colliding balls is not involved), there is no way to determine from the events depicted the direction in which the movie should be run.

Similarly, imagine a movie of a ball being thrown upward in the air and then caught again as it comes down. As the ball goes up, its initial kinetic energy is converted to potential energy. As it comes down, the reverse occurs— its potential energy is converted to kinetic energy. If we run the movie of this process backward, we again see a ball going up in the air and coming down. Nothing strange or unphysical is observed. If a process looks equally plausible in a movie run backward as in one run forward, it is **reversible.**

This is not accidental: the laws of physics (either newtonian or quantum) that describe the interactions of fundamental particles are completely time-symmetric. If a given such interaction is physically possible, then a movie of that interaction run in reverse shows an interaction that is also physically possible. There is no forward arrow of time implied by these laws of physics.

The behavior of macroscopic objects can be dramatically different. Consider the simple case of a box sliding across a level floor (see figure T1.1). As the box slides, friction slows it down, meaning that its kinetic energy is decreasing. If the floor is level, however, the box's potential energy is unchanged. Where does its kinetic energy then go? In unit C, we saw that the box's kinetic energy is converted to thermal energy, which consequently causes the temperature of the rubbing surfaces to increase.

Superficially, this looks like a simple energy transformation process similar to the case of the ball being thrown into the air; but consider a reversed movie of this process. The reversed movie shows a box sitting at rest on the floor that spontaneously accelerates while the rubbing surfaces get cooler. Any viewers unaware that the movie is reversed would really sit up in their seats, because we all know that the simple energy transfer process shown (thermal energy to kinetic energy) does *not* occur in nature, even though this conversion would be completely consistent with the conservation of energy. The conversion of kinetic energy to thermal energy through friction is **irreversible:** the time-reversed process is *not* physically possible.

My professor in an upper-level thermal physics class once handed each of several groups of us a movie camera and told us to make a movie illustrating some principles of thermal physics. One group of students (not mine, I regret to say) made their movie and then rewound it so that it would be shown in reverse. They cleverly intermixed reversible and irreversible processes so parts of the movie looked completely normal, while other parts were completely outrageous. The movie ended (began?) showing the paper

Irreversible processes

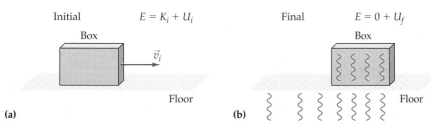

Initial $E = K_i + U_i$ Final $E = 0 + U_f$
       Box   Box

                    $\vec{v}_i$

                          Floor   Floor
(a)   **(b)**

Figure T1.1
The friction interaction between the floor and a sliding box converts kinetic energy to thermal energy. This process is *irreversible:* although the reverse process could be consistent with conservation of energy, it is never observed to occur.

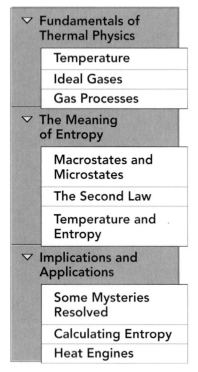

Figure T1.2
A chart illustrating how the chapters of unit T are organized into three subunits.

The great idea of the unit

The structure of the unit

cover of our thermal physics textbook being miraculously created from a heap of ashes in a fire. One of my favorite sequences began with the image of a placid pond on a beautiful spring day. Suddenly, strange ripples began to form on the pond's surface. Growing stronger, these ripples converged to a point on the pond, which suddenly disgorged a large rock. The rock leapt out of the pond into the air, falling into the hands of a surprised student. This movie very vividly made the point that some processes are just not physically possible in reverse.

Why is this so? This question becomes even more acute when we seek to explain this irreversible behavior in terms of the interactions between microscopic pointlike particles, which (as we've seen) are *reversible*. This seems logically absurd! How can reversible microscopic processes lead to irreversible macroscopic processes?

Ludwig Boltzmann was one of the first physicists to really understand that this contradiction was only apparent. During the 1870s, he published a series of fundamental papers that explained the relationship between the energy, temperature, and entropy of macroscopic objects and the microscopic motions of atoms in those objects. His work was energetically criticized by many physicists who doubted the reality of atoms and felt that physical theories should not be based on such unobservable and hypothetical entities. Many also dismissed his work out of hand because they could not see how irreversibility could be consistent with reversible microscopic interactions. As a result of struggling against this criticism, Boltzmann became increasingly despondent, finally committing suicide in 1906 just before experiments with brownian motion and quantization of electric charge made it completely clear that the atomic hypothesis was correct.

The "great idea" of this unit is really that Boltzmann was right: *the irreversible behavior of a complex system can be explained by the statistical consideration of the reversible interactions of its molecules.* Our goal in this unit is to develop the ideas and techniques we need to understand this basic idea.

A chart illustrating the organization of the unit appears in figure T1.2. This unit is divided into three subdivisions. The first (chapters T1 through T3) develops the basic concepts of *temperature, heat,* and *thermal energy* from both a macroscopic and a microscopic viewpoint, developing tools and ideas that we will use in the remainder of the unit. In chapters T4 through T6, we will explore the meaning of *entropy* (which is crucial for describing irreversibility) from the perspective of statistical mechanics. Chapters T7 through T9 discuss some of the applications and implications of the entropy concept, looking at the implications for microscopic systems in chapter T7, everyday processes in chapter T8, and heat engines in chapter T9.

Exercise T1X.1

Which of the following physical processes are (at least approximately) reversible and which are irreversible?

a.   A glass of milk spills on the floor.
b.   An object moves downward, compressing a spring.
c.   Ink poured into a glass of water gradually disperses throughout the water.
d.   A dropped book slams into the floor and remains at rest afterward.
e.   The moon orbits the earth.

T1.3   The Paradigmatic Thermal Process

The sliding box example vividly illustrates a process that converts energy irreversibly from one form to another. There is, however, another common process that in the long run will be more useful to us in clarifying the nature of irreversibility.

Imagine dropping a hot piece of metal into a beaker of cold water (see figure T1.3). We can observe (by touching the metal and water if necessary) that the metal gets cooler and the water gets warmer until their temperatures no longer appear to change. When the temperatures of the metal and of the water no longer discernibly change, we say the two objects are in **thermal equilibrium.** This process, like the sliding box example, is irreversible: one never sees a warm metal block placed in warm water spontaneously get hotter while the water becomes colder.

What is happening here? Colloquially, we might say, "Heat flows from the metal to the water." It certainly seems as if *something* is flowing out of the metal (leaving it cooler) and into the water (making it warmer) until equilibrium is established. In unit C, we learned what flows here is actually *energy:* as energy flows from the block to the water, the thermal energy of the block decreases (making it colder) while the thermal energy of the water increases (making it warmer).

This process will serve for us as the paradigm of an irreversible physical process. It is simple and commonplace, and yet it raises all the big questions that we will have to address in this unit:

1. *What is temperature?* What do we really mean when we say *hot* and *cold?* How is temperature related to thermal energy? How is temperature linked to other properties of an object?
2. *What is heat?* How is it related to temperature and internal energy? Why does heat spontaneously flow from a hot object to a cold object?
3. *What is thermal equilibrium?* Once energy does begin to flow between the metal and water, why does it stop? How is thermal equilibrium related to temperature?
4. *Why is the paradigmatic process irreversible?* Why does heat spontaneously flow from a hot object to a cold object but never the reverse?

The paradigmatic process

Questions raised by this paradigmatic process

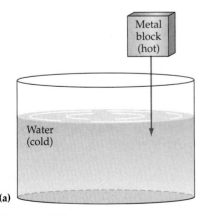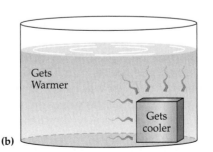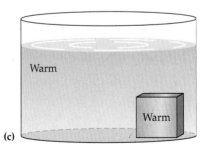

(a)   (b)   (c)

Figure T1.3

The paradigmatic thermal process. (a) A hot object (a block of metal here) is placed in contact with a cold object (water here). (b) The hot object gets cooler, and the cold object warmer (c) until their temperatures no longer change; at this point, the objects are in *thermal equilibrium.*

As we work through this unit, we will return again and again to this **paradigmatic thermal process** as we address the questions that it raises. The ideas we develop to explain this particularly simple process will help us to understand more complicated thermal processes as well.

T1.4   Temperature and Equilibrium

The first step in making a science out of thermal physics is to learn how to quantify an object's temperature in a meaningful and reproducible manner. But what *is* temperature, really? In this section and the next, we will create a meaningful quantitative definition of temperature from more primitive ideas.

What we colloquially call "temperature" is something we measure qualitatively by touch: we can feel when an object is *hot* and when it is *cold,* and we can even crudely distinguish degrees of the same. Measurement of temperature by touch is subjective and sometimes inconsistent. Even so, our crude direct sense of temperature allows us to discover that the physical characteristics of many objects change as their temperatures change. For example, as temperature increases, the electric resistance of a conductor changes, the pressure of a confined gas increases, certain kinds of liquid crystals undergo color changes, and almost all substances expand.

Quantifying temperature with a thermoscope

Imagine that we construct a device that measures a temperature-dependent property of something we can use as a probe and displays a quantitative result (e.g., an ohmmeter displaying a metal bar's resistance). Imagine further that we verify that (at least over some specified range) the value displayed always increases when we add thermal energy to the probe (say, with a flame) and decreases when we remove energy (say, by putting it in a freezer). We can then call such a device a **thermoscope.** A thermoscope is not a *thermometer,* because the value displayed is not *equal* to the probe's temperature: all we know is that the displayed value is *related* to its temperature (maybe in a complicated way) and that unique thermoscope values correspond to unique temperatures.

The link between temperature and equilibrium

When we bring a thermoscope probe into close physical contact with another object, we notice that the thermoscope display settles down to a fixed number as the probe and object come into equilibrium. Further experiments show that *two objects A and B will be in thermal equilibrium with each other if and only if the value displayed by a thermoscope T in equilibrium with each is the same.* For example, in our paradigmatic process, the values our thermoscope probe displays when we put it in successive contact with the hot metal block and the cold water are initially different but always become equal as the block and water come into equilibrium (see figure T1.4).

This is an empirical result: there is no reason that it *has* to be true. For example, if person T likes people A and B equally, it does not follow that A and B will even like each other at all. If an electrically charged object T equally attracts charged objects A and B, it does not follow that A and B will attract each other (in fact, they will repel). Thermoscope values *could* behave in a similar way, but experimentally they do not.

Since values displayed by a thermoscope are uniquely linked to the quantity that we want to call temperature, we can draw the following conclusion about temperature:

The zeroth law of thermodynamics

A well-defined quantity called **temperature** exists such that two objects in thermal contact will be in thermal equilibrium *if and only if* both have the same temperature.

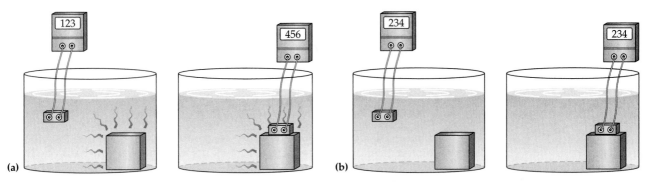

Figure T1.4
Imagine that we place a hot metal block into cold water. (a) Before they come into thermal equilibrium, they have different temperatures. (b) When the block and water come into thermal equilibrium, though, both are always found to have the *same* thermoscope reading.

This is called (somewhat whimsically) the **zeroth law of thermodynamics** because physicists now recognize it as being *logically* prior to laws that, when the zeroth law was first discussed, had already been named the "first" and "second" laws of thermodynamics.

This law in fact describes *the* essential physical characteristic of the quantity we call temperature. While most *apparent* thing about temperature is our sensation of hot and cold, from the perspective of physics, the thing that makes the idea of temperature meaningful and useful is that it *characterizes equilibrium.*

T1.5   Thermometers

As I stated earlier, a thermoscope does not display temperature directly. Values displayed by resistance thermoscopes using different bars of metal, for example, will be entirely different, even under the same conditions. How can we define a universal temperature scale that everyone can agree on?

We will later see that the zeroth law of thermodynamics, in conjunction with the concept of entropy, provides us with a means of *mathematically* defining temperature in terms of statistical ideas. But in the absence of this mathematical definition, the best we can do is to choose a particular thermoscope to *define* the temperature scale. A good thermoscope for this purpose will have the following characteristics:

Characteristics of a good standard thermoscope

1. It should be based on an easily measured property of a common substance.
2. The chosen property should monotonically increase in value with what our senses and other experiments tell us is "higher" temperature, as described previously.
3. The property should be measurable and satisfy the previous requirement over as wide a range of temperatures as possible.

A metal-bar thermoscope, for example, is a good thermoscope over a reasonable range of temperatures, but has several flaws that make it an unsuitable choice for the thermoscope that *defines* temperature. While it handily satisfies the first item of this list, at low temperatures, the resistance of a metal bar can actually decrease with increasing temperature, reach a minimum, and then increase again. This means that in this temperature range, we might measure the same resistance when the metal bar is in equilibrium with objects *A* and *B* separately and yet find that *A* and *B* are not in thermal equilibrium when we bring them together. A metal-bar thermoscope also obviously fails at temperatures above the metal's melting point.

Exercise T1X.2

Household thermometers typically measure the expansion of mercury or colored alcohol in a glass tube. Of the flaws listed above, which do you think is the main flaw of thermometers?

The constant-volume gas thermoscope

After a certain amount of turmoil and discussion, the physics community eventually settled on the **constant-volume gas thermoscope** as the standard thermoscope that quantitatively defines temperature. A simplified version of such a device is illustrated in figure T1.5. The temperature registered by such a thermoscope is *defined* to be proportional to the pressure *P* exerted by the gas when it is in thermal equilibrium with whatever object is being measured:

$$T \equiv CP \qquad \text{(where C is some constant of proportionality)} \qquad \text{(T1.1)}$$

The **pressure** *P* of a gas (or any fluid) is defined to be the *magnitude* of the force per unit area that it exerts on any surface that separates it from a vacuum. The SI unit of pressure is the **pascal** (abbreviated as Pa), where 1 Pa $\equiv$ 1 Newton per square meter (N/m^2). **Standard air pressure** has a defined value of 101.3 kilopascals (kPa); this is approximately equal to the average pressure of the earth's atmosphere at sea level.

In the hypothetical version of the constant-volume gas thermoscope shown in figure T1.5, a piston confines the gas in a cylindrical container. When we want to measure the temperature of an object, we place the cylinder containing the gas in thermal contact with that object and place weights on the piston until the volume of the gas has some specified value (indicated by marks on the cylinder walls). The pressure of the gas when it is confined to this exact volume is the total weight placed on the piston divided by the area of the piston (plus the ambient atmospheric pressure if the experiment is not conducted in a vacuum).

As a practical device capable of making precision measurements, this particular design has many flaws, but it nicely illustrates the *principle* of the

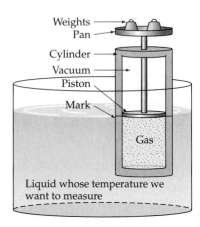

Figure T1.5

An idealized constant-volume gas thermoscope.

constant-volume thermoscope. (Problem T1R.2 discusses a more practical design.)

Having settled on a standard thermoscope, we need only set the constant of proportionality in equation T1.1 to determine the temperature scale. We can do this by choosing a value for the temperature of some well-defined and easily reproduced physical situation. Early investigators used the melting and boiling points of water as a standard reference, but it turns out that the exact temperatures of these points are sensitive to various kinds of conditions (including atmospheric pressure) and therefore are imprecisely defined. But it happens that at a certain precise temperature and pressure (0.61 kPa), water can coexist in equilibrium as solid, liquid, and gas simultaneously: this is called the **triple point** of water. This is a well-defined and easily reproducible physical reference point for temperature. By international agreement the triple point of water is defined to have a temperature of exactly 273.16 K, where the **kelvin** (abbreviated as K) is the SI unit of temperature. The temperature of any object is therefore defined to be

Defining the Kelvin temperature scale

$$T \equiv 273.16\,\text{K}\left(\frac{P}{P_{\text{TP}}}\right) \quad\Rightarrow\quad C \equiv \frac{273.16\,\text{K}}{P_{\text{TP}}} \qquad\text{(T1.2)}$$

where P is the pressure registered by a given gas thermoscope in close contact with the object and P_{TP} is the pressure registered by that same thermoscope at the triple point of water.

The temperature of the triple point was specifically chosen so that a difference in temperature of 1 K almost exactly corresponds to a difference in temperature of 1°C on the Celsius scale, which was the previous scientific standard. (Note that the kelvin is treated as any other unit in the SI: specifically, neither the word *degree* nor the degree symbol ° is used for temperatures in kelvins.)

Note also that the constant-volume gas thermoscope defines a temperature of 0 K (where the pressure of the gas is zero). This is an **absolute zero** in the sense that all physically measurable temperatures will be greater than zero (the pressure exerted by a gas can never be negative!). We will learn more about the physical meaning of *absolute zero* in the next chapter.

One of the reasons that the constant-volume gas thermoscope was chosen to be the standard is that the ratio of pressures P/P_{TP} in a given situation turns out to be fairly independent of the type and amount of gas used. Moreover, one finds experimentally that the small differences that actually *do* exist between gases get even smaller as we decrease the density ρ of the gas we use in our thermoscope (this also increases the range over which the thermoscope can be used by lowering the gas's boiling point). The *technical* definition of gas thermoscope temperature is thus

The zero-density limit

$$T \equiv 273.16\,\text{K}\left(\lim_{\rho\to 0}\frac{P}{P_{\text{TP}}}\right) \qquad\text{(T1.3)}$$

Purpose: This equation empirically defines an object's temperature T in kelvins.

Symbols: P is the gas pressure in a constant-volume gas thermoscope in equilibrium with the object, and P_{TP} is the pressure of the same thermoscope in equilibrium with water at its triple point.

The formal definition of
temperature

> **Limitations:**   The limit symbol indicates that we should evaluate
> the ratio in the limit that the density of gas in the thermoscope goes to
> zero. This equation defines temperatures on the Kelvin scale only.
>
> **Notes:**   This empirical definition of T will be superceded by a more
> fundamental definition we will discover in chapter T6.

If we use this definition, we do not need to specify the type of gas to use, nor does our gas sample need to be pure (we could even use ordinary air).

Experiments indicate that the gas thermoscope definition of temperature is fully consistent with the zeroth law. A gas thermoscope can also be used to register both extremely high temperatures (up to the melting point of the container) and extremely low temperatures (particularly if the gas is helium at very low density) or anywhere in between. An accurate constant-volume gas thermoscope is easily constructed from readily available materials. The fact that all gases yield the same ratio P/P_{TP} in the low-density limit when in contact with the same object implies that this ratio expresses something deep and fundamental about how temperature affects a gas that the physical characteristics of any *particular* gas only slightly modify. There are, therefore, many practical and theoretical reasons why the constant-volume gas thermoscope is an excellent standard.

Even so, I want to emphasize that this choice is nonetheless *conventional.* Another kind of thermoscope could (in principle) have been selected as the standard. It is therefore merely good fortune that this practical definition of temperature happens to coincide in a simple way with the more fundamental definition of temperature that we will discuss in chapter T6.

A thermometer is a calibrated
thermoscope

A **thermometer** is any thermoscope that has been calibrated so that when it is placed in contact with an object, it displays a temperature for that object that is *equal* to the extrapolated zero-density reading of a constant-volume gas thermoscope computed using equation T1.3. Thus, the distinction between a *thermoscope* and a *thermometer* is that the latter has been calibrated to be consistent with the accepted standard thermoscope.

Converting between
temperature scales

There are actually three different temperature scales in common use in the United States: the Kelvin scale, defined by equation T1.3; the older Celsius scale; and the still older Fahrenheit scale. These scales differ in the zero point of the scale and/or the size of the temperature unit. To convert from one scale to another, do the following:

1.   Express the temperature on each scale as a *temperature difference* from some common reference point.
2.   Multiply the difference on the original scale by a unit operator expressing the ratio between equivalent temperature differences on both scales to convert the temperature unit.
3.   Solve for the temperature on the final scale.

For example, water freezes at $0°C$ on the Celsius scale and at $32°F$ on the Fahrenheit scale, so we can use this as a common reference point for a conversion between the two. A temperature *difference* of $5°C$ is equivalent to $9°F$ (see exercise T1X.3), so to convert a temperature $T_{[C]}$ on the Celsius scale to the equivalent temperature $T_{[F]}$ on the Fahrenheit scale, we do the following:

$$\left(T_{[F]} - 32°F\right) = \frac{9°F}{5°C}\left(T_{[C]} - 0°C\right) \quad \Rightarrow \quad T_{[F]} = \frac{9°F}{5°C}T_{[C]} + 32°F \qquad (T1.4)$$

Note that in the first expression, the quantities in parentheses are temperature differences on each scale from the reference point, and in the last step, I have solved for $T_{[F]}$.

Exercise T1X.3

Given that the boiling point of water is 100°C and 212°F, show that a temperature difference of 5°C is equivalent to 9°F.

Similarly, the freezing point of water is 0°C but 273.15 K, and the Kelvin scale is defined so that a temperature difference of 1°C is the same as a difference of 1 K. So to convert from a temperature $T_{[C]}$ on the Celsius scale to a temperature T on the Kelvin scale, we do the following:

$$(T - 273.15\,\text{K}) = \frac{1\,\text{K}}{1°C}\left(T_{[C]} - 0°C\right) \quad \Rightarrow \quad T = \frac{1\,\text{K}}{1°C}T_{[C]} + 273.15\,\text{K} \quad (T1.5)$$

(Note that in this unit, you should consider T *without* a unit subscript in brackets always to refer to a temperature on the Kelvin scale.)

You can use the same approach to develop conversion formulas for other temperature scale pairings. Some "benchmark" temperatures on the Kelvin, Celsius, and Fahrenheit scales are listed in table T1.1.

Exercise T1X.4

What is the formula for converting a Fahrenheit temperature to a Celsius temperature?

Table T1.1   Selected temperature benchmarks

Center of the sun	1.5×10^7 K	1.5×10^7°C	2.7×10^7°F
Surface of the sun	5800 K	5500°C	10,000°F
Melting point of tungsten	3683 K	3410°C	6170°F
Melting point of iron	1808 K	1535°C	2795°F
Melting point of lead	601 K	328°C	622°F
Boiling point of water	373 K	100°C	212°F
Normal body temperature	310 K	37°C	98°F
Room temperature	295 K	22°C	72°F
Freezing point of water	273 K	0°C	32°F
Boiling point of nitrogen	77 K	−196°C	−321°F
Boiling point of helium	4.2 K	−269°C	−452°F
Background temperature of universe	2.7 K	−270.5°C	−454.8°F
Lowest laboratory temperatures	<0.1 μK	−273.15°C	−459.7°F

Exercise T1X.5

On a cold Minnesota winter day, the temperature is −40°F. What is this temperature on the Celsius scale? On the Kelvin scale?

Figure T1.6
A cold day in Minnesota.

Note that the size of the temperature unit we use is not physically significant, but the zero point of the scale *is*. We will find in subsequent chapters that a number of physical quantities depend on temperature in a simple way only when we measure temperature from absolute zero. We will therefore use the Kelvin scale almost exclusively in this text.

T1.6   Temperature and Thermal Energy

Temperature is linked to thermal energy

In unit C, we saw that there was a connection between an object's temperature and its internal thermal energy. Empirically, we found that under normal circumstances, if we convert a given amount of kinetic energy to thermal energy in an object, its temperature *increases* by an amount given by the equation

$$dU = mc\,dT \qquad \text{or} \qquad c \equiv \frac{1}{m}\frac{dU}{dT} \qquad (T1.6)$$

Purpose: This equation describes the empirically observed link between the change dU in an object's internal energy and its corresponding change in temperature dT.

> **Symbols:** m is the object's mass, and c is the object's **specific heat** (*not* the speed of light!).
> **Limitations:** dU and dT must be small enough that $c \approx$ constant. The value of c is often undefined during a phase change.
> **Notes:** The second version of this equation amounts to a definition of c. *Specific heat*, though conventional, is a horribly confusing misnomer here, because we can add energy in forms other than heat. (*Specific* means "per unit mass.")

This equation is useful because the quantity c so defined empirically is *strongly* dependent on an object's composition and phase, but *independent* of its size and only *weakly* dependent on its temperature. Values of c range from 4186 joules per kilogram per kelvin (abbreviated as $J\ kg^{-1}\ K^{-1}$) (for liquid water) to $127\ J\ kg^{-1}\ K^{-1}$ (for solid lead) at normal temperatures. You might review the examples in unit C (or look at problem T1S.8) to see how we can use this equation to determine equilibrium temperatures.

By the way, in this unit, U will refer *exclusively* to an object's *thermal energy* unless otherwise stated.

The idea that thermal energy is indeed *energy* proved very helpful to us in unit C, making it possible for us to explain how energy is conserved in situations where it superficially seems to disappear. In our present context, however, this concept raises some perplexing questions:

1. Exactly how is energy stored in an object?
2. Why and how is an object's *temperature* linked to its thermal energy?
3. Why does energy flow spontaneously from high to low temperature (as in our paradigmatic thermal process) but never the other way around?

We discussed the first question to some degree in unit C, but the others only really arise in the context of this chapter. In this chapter, we have defined temperature in terms of the pressure in a gas thermometer and linked it to equilibrium. What has *either* of these ideas to do with thermal energy?

In the next chapter, we will begin to address the first two questions by using a simple microscopic model of a gas: we will see exactly how a gas stores energy and how this energy is related to temperature. The third question is a special case of the general problem of irreversibility, which we will begin to address in chapter T4. In chapter T6 we will finally see how temperature, energy, and entropy are all linked to the concept of equilibrium, providing full and satisfying answers to these perplexing questions.

U means "thermal energy"

Questions raised by the link between U and T

TWO-MINUTE PROBLEMS

Note: The answers to problems T1T.1 through T1T.3 are deliberately debatable!

T1T.1 Characterize each of the following processes as being reversible (A) or irreversible (B).
  a. A living creature grows.
  b. A ball is dropped and falls freely downward.
  c. A ball rebounds elastically from a wall.
  d. A piece of hamburger meat cooks on a grill.
  e. A cube of ice melts in a glass.
  f. A bowling ball elastically scatters some bowling pins.

T1T.2 All irreversible processes involve macroscopic objects [true (T) or false (F)] and always involve transfers to the thermal energy of that object (T or F).

T1T.3 Certain kinds of liquid crystals change color as their temperatures change. A sheet with a layer of such crystals is (A) a thermometer or (B) a thermoscope or (C) neither. (If the sheet is not a thermometer, what would you have to do to make it a thermometer?)

T1T.4 Imagine that we place an aluminum cylinder, a wooden block, and a Styrofoam cup on a table and leave them there for several hours. We then come back into the room and feel each object. Which (if any) feels coolest? Which (if any) *actually* is coolest?
A. The aluminum cylinder
B. The wooden block
C. The Styrofoam cup
D. All are the same

T1T.5 Is pressure a (A) *vector* or (B) *scalar* quantity?

T1T.6 Imagine that we place objects A and B into a large bucket of water and allow them to come into equilibrium with the water. If we now extract A and B from the water and immediately place them in contact with each other, they will necessarily be in equilibrium with each other (T or F).

T1T.7 Which of the following equations is the correct equation for converting a temperature in kelvins to the equivalent temperature on the Fahrenheit scale?

A. $T_{[F]} = \left(\dfrac{9\,°F}{1\,K}\right)(T)$

B. $T_{[F]} = \left(\dfrac{9\,°F}{5\,K}\right)(T - 32\,°F) + 273.15\,K$

C. $T_{[F]} = \left(\dfrac{5\,K}{9\,°F}\right)(T - 32\,°F) + 273.15\,K$

D. $T_{[F]} = \left(\dfrac{9\,°F}{5\,K}\right)(T - 273.15\,K) + 32\,°F$

E. $T_{[F]} = \left(\dfrac{9\,°F}{5\,K}\right)(T - 273.15\,K) - 32\,°F$

F. Other (specify)

T1T.8 Imagine that we place a 100-gram (100-g) aluminum block with an initial temperature of 100°C in a Styrofoam cup containing a 100-g sample of water at 0°C. (The specific heats of aluminum and water are 900 J kg^{-1} K^{-1} and 4186 J kg^{-1} K^{-1}, respectively.) The final temperature of the system will be closest to (A) 0°C, (B) 20°C, (C) 50°C, (D) 80°C, or (E) 100°C.

HOMEWORK PROBLEMS

Basic Skills

T1B.1 In figure T1.5, how much weight will we have to put on the pan if the pressure of the gas is 90 kPa and the area of the piston is 3.0 cm^2? (Ignore the weight of piston itself, and assume we are conducting the experiment in a vacuum.)

T1B.2 A piston with an area of 1.5 cm^2 slides freely in the middle of a sealed cylinder, as shown below. The pressure of the gas on the left is 150 kPa, while the pressure of the gas on the right is 120 kPa. What are the magnitude and direction of the force you would have to manually apply to the piston to keep it at rest?

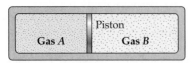

T1B.3 Imagine that we measure the pressure of the helium gas in a constant-volume gas thermoscope to be

32.0 kPa at the triple point of water and 42.3 kPa when immersed in a certain liquid. What is the temperature of that liquid?

T1B.4 Imagine that we measure the pressure of the helium gas in a constant-volume gas thermoscope to be 55.0 kPa at the triple point of water and 42.3 kPa when immersed in a certain liquid. What is the temperature of that liquid?

T1B.5 What is the equation for converting a temperature on the Fahrenheit scale to the equivalent temperature in kelvins?

T1B.6 The hottest recorded daytime temperature on earth is about 130°F. What is this temperature on the Celsius scale? In kelvins?

T1B.7 The lowest officially recorded temperature within the continental United States is about −70°F. What is this temperature on the Celsius scale? In kelvins?

T1B.8 Body temperature is often quoted as being 98.6°F. However, a given person's body temperatures vary over a range of as much as $\pm 1°$F during a day, and can vary by about as much from person to person. So it seems strange to quote this temperature to a tenth of a degree: it would make more sense to round it to the nearest integer and say that it is 99°F. However, convert 98.6°F to the equivalent temperature on the Celsius scale. Given that *most* European scientists in the 19th century were using the Celsius scale, speculate on why body temperature was stated as 98.6°F instead of 99°F in countries using the Fahrenheit scale.

Synthetic

T1S.1 Your lab partner claims that physicists have it all backward: cold things actually have more thermal energy than hot things, and energy actually flows from cold to hot. What evidence could you point out that would contradict this assertion?

T1S.2 Do we have to represent hotter temperatures by higher numbers, or is this just a convention? If it is not a convention, explain why hotter temperatures *necessarily* must be represented by higher numbers. If this is a convention, suggest why people might have been prompted to choose this convention.

T1S.3 The pressure in a body of water increases with depth. To see exactly *how* the pressure increases, consider the diagram shown below. Imagine that the water is initially at rest. We then surround a certain block of water with an imaginary rigid plastic cylinder of essentially zero weight, isolating the water inside the cylinder from that outside. Simply isolating the water in this way should not cause it to move: if it was at rest before being enclosed, it will remain at rest. If it *is* at rest, though, the net force on the cylinder must be zero. Since the water's gravitational interaction with the earth exerts a downward force on the cylinder equal to the weight of the water inside the cylinder, there must be some other force pushing the water upward. Thus the water pressure pushing upward on the bottom face of the cylinder must be greater than the water pressure pushing downward on the top face. If the area of each cylindrical face is A, then the upward force on the lower face has a magnitude of $P_1 A$ (where P_1 is the pressure at vertical position z_1) and the downward force on the upper face has a magnitude of $P_2 A$ (where P_2 is the pressure at vertical position z_2). Argue that for the water in the cylinder to remain stationary, P_2 and P_1 have to be related as follows

$$P_2 - P_1 = -\rho g(z_2 - z_1) \qquad (\text{T1.7})$$

where ρ is the density of water. This equation implies that the magnitude of the change in pressure is proportional to the magnitude of the change in depth (assuming ρ is independent of depth).

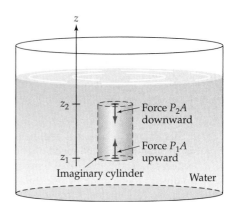

T1S.4 The pressure at the top of a body of water on the surface of the earth is simply the atmospheric pressure P_a there. At what depth below the surface is the water pressure on your body $P = 2P_a$? (*Hint:* Consider problem T1S.3. Assume that $P_a = 101.3$ kPa and $\rho = 1000$ kg/m^3.)

T1S.5 (a) *Argue*, using equation T1.7, that if the pressure at the top surface of a fluid is zero and its density ρ is constant, the downward pressure on a horizontal surface of area A at the bottom of the fluid is equal to W/A, where W is the magnitude of the weight of the fluid enclosed by an imaginary vertical column whose base has area A.
(b) How thick would the earth's atmosphere be if the density of air did not vary with altitude? (Assume that the pressure and density of air at the earth's surface are 101 kPa and about 1.3 kg/m^3, respectively.)

T1S.6 Use equation T1.7 to prove that if the pressure at the top surface of a fluid is zero, the downward pressure on a horizontal surface of area A at the bottom of the fluid is equal to W/A, where W is the magnitude of the weight of the fluid enclosed by an imaginary vertical column whose base has area A *even if the fluid's density ρ depends on vertical position*. (*Hint:* Argue that equation T1.7 still applies if the change in vertical position is infinitesimal. Integrate to find the total pressure difference between the top and bottom of the fluid, and do something similar to find the total weight of the fluid. You should find that you do not even need to assume that g is independent of vertical position.)

T1S.7 The Rankine temperature scale [where temperatures are expressed in degrees Rankine (°R)] is defined so that temperature differences in °R are the same as those in °F, but the zero of the Rankine scale is at absolute zero.
(a) Find conversion formulas between temperatures on the Rankine scale and those on the Fahrenheit and Kelvin scales.
(b) What is room temperature on the Rankine scale?

T1S.8 Imagine that we place object A with high initial temperature T_A in contact with object B with low initial temperature T_B. We know from experience that thermal energy will flow from A to B under these circumstances. If these objects are isolated from everything else during this process, then any energy lost by A will be gained by B:

$$dU_A = -dU_B \tag{T1.8}$$

(a) If this change of thermal energy is "sufficiently small" that the objects' specific heats are approximately constant over the temperature ranges involved, then show, using this equation and equation T1.6, that

$$-\frac{dT_B}{dT_A} = \frac{m_A c_A}{m_B c_B} \tag{T1.9}$$

(b) Now, since both objects eventually come to the same final temperature T_f, the total change in the temperature of A is $dT_A = T_f - T_A$ and that for B is $dT_B = T_f - T_B$. To save writing, let us define $u \equiv m_A c_A / m_B c_B$. Plug these relations into equation T1.9, and show (after some algebra) that

$$T_f = T_B + \frac{u}{1+u}(T_A - T_B) \tag{T1.10}$$

(c) Note that if object A is much more massive than object B, then u will be very large and $u/(1+u) \approx 1$, meaning that $T_f \approx T_B + T_A - T_B$, as one might expect. Argue that equation T1.10 also gives a plausible result if object B is much more massive than object A.

T1S.9 Imagine that we put a 100-g block of aluminum with an initial temperature of 100°C into a cup containing 250 g of water with an initial temperature of 25°C. What is the final equilibrium temperature of this system? (*Hint:* You can use the result of problem T1S.8. The specific heats of aluminum and water are 900 J kg⁻¹ K⁻¹ and 4186 J kg⁻¹ K⁻¹, respectively.)

T1S.10 Imagine that we put a 150-g steel ball with an initial temperature of 0°C into a cup containing 150 g of water with an initial temperature of 100°C. What is the final equilibrium temperature of this system? (*Hint:* You can use the result of problem T1S.8. The specific heats of iron and water are 450 J kg⁻¹ K⁻¹ and 4186 J kg⁻¹ K⁻¹, respectively.)

Rich-Context

T1R.1 Consider the situation shown below. Argue that the pressure on the top side of the card times the cross-sectional area of the tube has to be equal to the weight of the water in the tube. The earth's atmosphere

exerts pressure on the bottom side of this card. If the latter is to hold the card tight against the bottom of the tube, what is the maximum possible height of water in the tube?

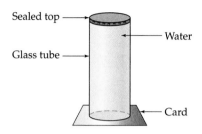

Sealed top — Water
Glass tube —
— Card

T1R.2 An alternative design for a constant-volume gas thermoscope is shown below.
(a) Find an expression for the gas pressure P in terms of the atmospheric pressure P_a, the density of mercury ρ and the height difference h between the mercury in the tubes, and/or the cross-sectional area A of the tubes.
(b) Describe the advantages of this design over the design shown in figure T1.7 (list as many advantages as you can).

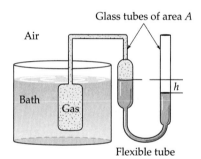

Glass tubes of area A
Air
Bath
Gas
h
Flexible tube

Figure T1.7
To measure the bath's temperature, we can raise or lower the tube on the right until the mercury (dark gray fluid) is exactly at the level of the mark. The gas pressure can then be found from h.

Advanced

T1A.1 Imagine that the density of air varies exponentially with altitude according to the following formula

$$\rho(z) = \rho(0)e^{-z/a} \tag{T1.11}$$

Find a value of a that gives the correct air pressure at the surface of the earth. (*Hint:* Study problem T1S.6.)

ANSWERS TO EXERCISES

T1X.1 Reversible: (b), (e); irreversible: (a), (c), (d).

T1X.2 The most serious problem with both mercury and alcohol thermometers is their limited range. Mercury freezes at $-39°C$, which is higher than possible wintertime temperatures in some parts of the United States. Methanol does not freeze until the temperature falls to $-94°C$, but boils at $65°C$, well below the boiling point of water. Thermometers constructed using these working fluids will thus be useful over only a limited range of temperatures.

T1X.3 The temperature difference between the freezing and boiling points of water is $100°C$ on the Celsius scale and $212°F - 32°F = 180°F$ on the Fahrenheit scale.

T1X.4 $T_{[C]} = \dfrac{5°C}{9°F}\left(T_{[F]} - 32°F\right)$

T1X.5 $-40°F = -40°C = 233\,K$

T2

Ideal Gases

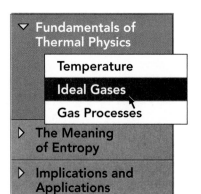

Chapter Overview

Introduction

In this chapter, we will explore a simple newtonian ideal gas model that will give greater insight into how a substance's temperature is related to its thermal energy and its other macroscopic properties.

Section T2.1: A Microscopic Model of an Ideal Gas

Our gas model assumes that a gas consists of a huge number of noninteracting molecules that are tiny compared to their separation. According to this model, the pressure that a gas exerts on a piston confining it to a cylinder is due to molecules elastically colliding with the piston as they move randomly inside the cylinder. This model implies mathematically that a gas's pressure P, volume V, and number of molecules N are linked to its average molecular x-velocity v_x by $PV/N = [mv_x]_{\text{avg}}$. Since $P \propto T$ at fixed V and N by the definition of temperature,

$$[mv_x^2]_{\text{avg}} = k_B T \qquad \text{(T2.6)}$$

The constant k_B is empirically independent of V and T for low-density gases, so

$$PV = Nk_B T \qquad \text{(T2.9)}$$

Purpose: This equation specifies the pressure P of N molecules of a low-density gas enclosed in a volume V at temperature T.

Symbols: k_B is **Boltzmann's constant** $= 1.38 \times 10^{-23}$ J/K.

Limitations: This equation strictly applies only in the zero-density limit (although it is a good approximation for real gases at typical densities). All parts of a gas sample must be in equilibrium for it to have a well-defined single value of P and of T throughout the volume (as the equation assumes).

Notes: We call this equation the **ideal gas law,** and any gas obeying this law is an **ideal gas.**

Section T2.2: Temperature and Energy

Since the coordinate directions are equivalent in a gas in equilibrium, equation T2.6 implies that $[mv^2]_{\text{avg}} = [mv_x^2]_{\text{avg}} + [mv_y^2]_{\text{avg}} + [mv_z^2]_{\text{avg}} = 3k_B T$, or

$$K_{\text{avg}} = \frac{1}{2}[mv^2]_{\text{avg}} = \frac{3}{2}k_B T \qquad \text{(T2.18)}$$

Purpose: This equation links the average molecular kinetic energy K_{avg} in an ideal gas to the gas's absolute temperature T.

Symbols: k_B is Boltzmann's constant.

Limitations: This equation applies when the ideal gas model is a good approximation.

Section T2.3: Molecular Speeds and Brownian Motion

Equation T2.18 implies that

$$v_{rms} \equiv \sqrt{[v^2]_{avg}} = \sqrt{\frac{3k_B T}{m}} \qquad \text{(T2.20)}$$

Purpose: This equation specifies the **root-mean-square, or rms, speed** v_{rms} of a gas molecule with mass m in a gas with absolute temperature T.
Symbols: k_B is Boltzmann's constant.
Limitations: This equation strictly applies only to an ideal gas.

This applies *roughly* to molecules in a liquid as well. Small particles such as pollen grains suspended in water can be visibly jostled by collisions with water molecules, a phenomenon known as **brownian motion.**

Section T2.4: The Thermal Energy of a Gas

One might expect from equation T2.18 that a gas's total thermal energy would be $U = Nk_{avg} = (3/2)Nk_B T$, but empirically

$$U = \frac{f}{2} Nk_B T \qquad \text{(T2.23)}$$

Purpose: This equation specifies the total thermal energy U of a sample of ideal gas containing N molecules at absolute temperature T.
Symbols: k_B is Boltzmann's constant. Near room temperature, $f \approx 3$ for purely **monatomic** gases, $f \approx 5$ for purely **diatomic** gases, and $f > 6$ for **polyatomic** gases. (Physicists call f the number of molecular *degrees of freedom* for reasons beyond our scope here.)
Limitations: This equation only applies to real gases in the ideal gas limit. The variable f is never exactly an integer, but it is very close to being an integer for monatomic and diatomic gases in certain temperature ranges.

This is true because diatomic (two-atom) and polyatomic (multiple-atom) gas molecules can store rotational and vibrational energy as well as their center-of-mass kinetic energy. However, we observe empirically that these energy storage modes appear to "switch on" only above certain temperatures.

Section T2.5: Solids and Liquids

It is difficult to build an analogous simple model for a liquid, but empirically, liquids also store energies that are a small multiple of $k_B T$ per molecule. Empirical measurements also suggest that water molecules, like gas molecules, have an average center-of-mass kinetic energy of $(3/2)k_B T$.

The thermal energy of a monatomic solid is observed to be fairly nearly $U = 3Nk_B T$ at room temperature, independent of the element involved (this is the **law of Dulong and Petit**). We will explain this result in chapter T6.

Section T2.6: Conclusions

We are left with some interesting questions: Why do the extra energy storage modes in diatomic and polyatomic molecules only switch on above certain temperatures, and store an energy of roughly $k_B T$ per molecule when they do? Why do monatomic solids appear to store almost exactly $3k_B T$ of energy per molecule? We will solve these mysteries in chapter T7.

T2.1   A Microscopic Model of an Ideal Gas

The goal of this chapter is to help you develop a deeper understanding of how thermal energy is related to temperature and how these macroscopic properties are connected with the microscopic behavior of molecules. In this section, I will propose a simple model for a gas that will provide a first step in making this connection clear.

In his *Hydrodynamica* (published in 1738), the Swiss physicist Daniel Bernoulli proposed that the pressure a gas exerts on a piston might be due to many tiny gas particles hitting the piston's inner surface. This hypothesis also offered an explanation for the extreme compressibility of gases (compared to liquids or solids), which suggests that there must be a lot of empty space between whatever particles of matter the gas contains. Bernoulli's suggestion was the first step toward the model I am about to describe.

As in the case of models that we have considered previously in this course, this model is a simplified picture that makes it easier to visualize and think about a given physical phenomenon but that still retains its crucial features. Since the model is simplified, the results that it predicts are likely to be only approximately true; but if the model is good, the approximation will be good.

Assumptions of the model

The model of the ideal gas that we will use in this section is based on six *assumptions*. In each case, I will describe the assumption and comment on the degree to which it is an approximation to real gases.

1. **A gas consists of a huge number of tiny molecules.** This is an excellent assumption: laboratory-sized samples of gas contain on the order of magnitude of 10^{23} molecules, an outrageously large number. Even a bacterium-sized sample of gas contains roughly one molecule for every person in the United States.
2. **Individual molecules are tiny compared to their average separation.** We will essentially treat the molecules as point particles in this model. This assumption is a fairly good description of real gases: the total volume of the molecules in a typical gas is about one-thousandth of the total volume of the gas, implying that most of the gas is empty space. This approximation clearly improves as the density of the gas decreases.
3. **The molecules obey Newton's laws of motion.** Superficially, this seems like a bad assumption: quantum effects (as discussed in unit Q) are often important for things as small as molecules. It turns out that in this case, however, since our derivations primarily involve momentum and energy conservation (principles that transcend all theories), the quantum-mechanical treatment of gases yields essentially the same results as a newtonian treatment. The newtonian model, on the other hand, is *much* easier to understand.
4. **The molecules don't interact with one another.** For certain kinds of gases (notably helium and other noble gases) this assumption is fairly accurate for a wide range of temperatures and densities. In other kinds of gases (notably water vapor) there can be significant interactions between molecules when they are close together: it is these interactions that allow the molecules to form a liquid or solid. The strengths of such interactions fall off sharply with distance, though, so this assumption accurately applies to any gas whose density is low enough that its molecules spend little time close to their neighbors.
5. **Collisions between the molecules and the container walls are elastic.** Like assumption 3, this assumption is not really true; but creating a

model where collisions are *not* assumed to be elastic involves much greater effort and yet yields the same results. The basic reason is that even though sometimes a molecule does transfer energy to the wall in a collision (or vice versa), when everything is in equilibrium, the *net* energy transferred to or from the wall is zero; so on average the molecules act *as if* the collisions were elastic. It is easier just to start with this assumption.

6. **The motion of the molecules is entirely random.** If the gas as a whole is at rest in its container (and is not sloshing around), this is an excellent approximation. This assumption means that we are just as likely to find a molecule moving in the $+x$ as in the $-x$ direction, and there is no intrinsic difference between the x, y, and z directions in the gas.

To summarize, assumptions 1 and 6 seem to be well founded as far as we know; assumptions 3 and 5 are not particularly sound, but a more rigorous analysis yields the same results; but assumptions 2 and 4 are substantial simplifications that will mean that predictions based on this "ideal gas model" may not accurately reflect real gases under certain circumstances. However, both assumptions 2 and 4 become more accurate as the density of the gas decreases.

Now let's see what this model predicts about the behavior of gases. If the pressure exerted by a gas on its container walls is really due to gas molecules bouncing off those walls, how can we compute the force exerted by these molecules? We can answer this question with the help of a "thought" experiment. Imagine that we confine a sample of gas in a cylinder by a piston that is free to move frictionlessly in response to the gas molecules hitting it, as shown in figure T2.1. Each collision interaction between a molecule and the piston delivers a certain tiny impulse (momentum transfer) to the piston. By the definition of force, the total average force that the gas exerts on the piston is the average rate at which these collisions transfer momentum to the piston

$$[\vec{F}]_{avg} \equiv \left[\frac{d\vec{p}}{dt}\right]_{avg} \qquad (T2.1)$$

We can keep the piston from moving in response to this force by exerting on it an opposing force of equal magnitude; in what follows we will assume that we do this.

For the sake of simplicity, let us focus on a *single* gas molecule of mass m bouncing around the chamber. Each time it hits the piston, we imagine that it rebounds elastically (in accordance with assumption 5), which means that kinetic energy is conserved in the collision. Since we are now holding the

The pressure exerted by a single molecule

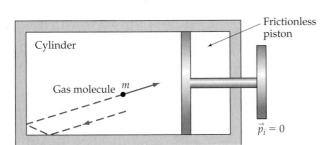 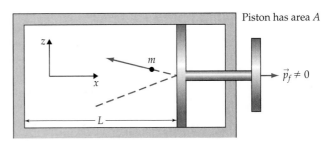

Figure T2.1
When a molecule hits an otherwise isolated piston, the molecule transfers an impulse to the piston. The total force that the gas exerts on the piston is the rate at which such collisions transfer momentum to the piston. (In practice, we will exert an external force on the piston to hold it at rest against the force exerted by the gas.)

piston at rest, conservation of kinetic energy here implies that the *molecule's* kinetic energy, and thus its speed, remains the same. In such a case, the collision simply reverses the component of the molecule's velocity perpendicular to the piston's face (v_x in the case shown in figure T2.1) without changing its magnitude or changing the other components of the molecule's velocity. The change in the molecule's x-momentum during a collision is thus

$$p_{x,\text{final}} - p_{x,\text{initial}} = -m|v_x| - m|v_x| = -2m|v_x| \qquad \text{(T2.2)}$$

Conservation of momentum thus implies that such a collision delivers an x-impulse $[dp_x] = +2m|v_x|$ to the piston.

Now, every time the molecule collides with a wall of the cylinder, either its x-velocity is unaffected (as when the molecule hits an upper or lower wall) or it is reversed without changing magnitude (as when the molecule hits the piston or the opposite wall). In any case, the value of $|v_x|$ is *not* changed by a collision with any wall. Since there are no forces acting on the molecule *between* such collisions by assumption 4, $|v_x|$ *never* changes. This means that the time interval between collisions of this particular molecule with the piston is $2L/|v_x|$, since the molecule must cover a distance of $2L$ in the x-direction (forward and back across the cylinder) between collisions with the piston. Therefore, the magnitude of the average force that this particular molecule exerts on the piston is

$$[F]_{\text{avg}} = |F_x|_{\text{avg}} = \frac{\text{impulse per collision}}{\text{time between collisions}} = \frac{2m|v_x|}{2L/|v_x|} = \frac{mv_x^2}{L} \qquad \text{(T2.3)}$$

The pressure on the piston is defined to be the magnitude of the average force acting on it divided by the area of its face. Dividing both sides of equation T2.3 by the piston's area A, we get

$$P \text{ (for one molecule)} \equiv \frac{[F]_{\text{avg}}}{A} = \frac{mv_x^2}{AL} = \frac{mv_x^2}{V} \qquad \text{(T2.4)}$$

since $AL = V$ is the cylinder's volume.

Exercise T2X.1

Explain *physically* why the pressure depends on the square of the molecular x-velocity. If the molecule's x-velocity doubles, why should the pressure quadruple?

The pressure exerted by N identical molecules

If the molecules do not significantly interact with one another (assumption 4), then the pressure exerted by N molecules will simply be the sum of the pressures due to each individual molecule. But the molecules do not necessarily move at the same speed as they rattle around the cylinder (we will see how molecular speeds are distributed in chapter T7). This means we cannot simply multiply equation T2.4 by N: we actually have to do the sum. Let P_i be the pressure due to the ith molecule, and let $v_{i,x}$ be that molecule's x-velocity and m_i be its mass. Then

$$P(\text{total}) \equiv \sum_{i=1}^{N} P_i = \sum_{i=1}^{N} \frac{m_i v_{i,x}^2}{V} = \frac{N}{V}\left(\frac{1}{N}\sum_{i=1}^{N} m_i v_{i,x}^2\right) \equiv \frac{N}{V}\left[mv_x^2\right]_{\text{avg}} \qquad \text{(T2.5)}$$

by definition of the average.

Now, the *definition* of temperature implies that for a fixed number of gas molecules N in a given fixed volume V, the pressure is proportional to its absolute temperature T. We can see from equation T2.5 that this will be true if and only if

$$[mv_x^2]_{\text{avg}} = k_B T \tag{T2.6}$$

where k_B is a constant of proportionality that we will call **Boltzmann's constant.** If we plug this back into equation T2.5, we find that

$$P = \frac{N}{V} k_B T \tag{T2.7}$$

Our argument here does not *require* that k_B be independent of N and V: all that we know is that P is proportional to T for all (sufficiently low-density) gases at *fixed* values of N and V. However, there is also no obvious physical reason why it *should* depend on N and V, so the simplest hypothesis is that k_B is simply a constant. Experimental measurements of $k_B = PV/NT$ for all gases at various values of P, N, V, and T in fact (in the low-density limit) always seem to yield the same numerical result

$$k_B = 1.38 \times 10^{-23} \text{ J/K} \tag{T2.8}$$

in strong support of this hypothesis.

We see that our simple gas model, coupled with this last experimental result, predicts that all gases (at least in the low-density limit) will obey equation T2.7. This equation is typically written in the form

The ideal gas law

$$PV = Nk_B T \tag{T2.9}$$

Purpose: This equation specifies the pressure P of N molecules of a low-density gas enclosed in a volume V at temperature T.

Symbols: k_B is Boltzmann's constant $= 1.38 \times 10^{-23}$ J/K.

Limitations: This equation strictly applies only in the zero-density limit (although it is a good approximation for real gases at typical densities). All parts of a gas sample must be in equilibrium for it to have well-defined single values of P and T throughout the volume (as the equation assumes).

Notes: We call this equation the **ideal gas law,** and any gas obeying this law is an **ideal gas.**

An *ideal gas* is an idealized model that only approximately models real gases. The ideal gas law in a sense describes the general abstract behavior of gases, behavior that is slightly perturbed by the specific characteristics of a given real gas. The densities of gases at atmospheric pressure and room temperature are generally low enough that deviations from the ideal gas law are tolerably small (although measurable).

Chemists usually write the ideal gas law in the form

$$PV = nRT \tag{T2.10}$$

where (using $N_A \equiv$ Avogadro's number $= 6.02 \times 10^{23}$ molecules per mole)

$$n \equiv \frac{N}{N_A} = \text{number of moles of gas in question} \tag{T2.11a}$$

$$R \equiv N_A k_B = \left(\frac{6.02 \times 10^{23}}{1 \text{ mol}}\right)\left(\frac{1.38 \times 10^{-23} \text{ J}}{1 \text{ K}}\right) = \frac{8.31 \text{ J}}{1 \text{ mol} \cdot \text{K}} \tag{T2.11b}$$

(The constant R is often called the **gas constant.**) A modest advantage of the chemist's form of the law is that n can easily be determined from macroscopic measurements

$$n = \frac{M}{M_A} \tag{T2.12}$$

where M is the mass of the sample and M_A is the gas's **molar mass** (sometimes incorrectly called its *molar weight*), defined to be the mass of a mole (i.e., Avogadro's number of molecules) of the gas, typically expressed in units of grams per mole (g/mol). In turn M_A can be quickly calculated by summing the atomic masses (which one can find listed in any periodic table) of all atoms appearing in the gas molecule. The value of M_A in grams per mole is also *approximately* the total number of nucleons (protons and neutrons) in the molecule's atoms. For example, a nitrogen molecule contains 2 nitrogen atoms, each with 7 protons and 7 neutrons, so the molecular mass of nitrogen is approximately $2 \times 14 = 28$ g/mol.

In this unit, though, we want to explain gas properties in terms of the *microscopic* behaviors of its molecules, so the physicist's version of the law (with its explicit reference to the number of molecules N) will be generally more meaningful, and I will use it consistently throughout the unit. Note that it is still pretty easy to calculate N for a given sample of gas:

$$N = N_A n = \frac{N_A}{M_A} M \tag{T2.13}$$

Do not confuse n (the number of moles) with N (the number of molecules).

Exercise T2X.2

Imagine that you have a bottle containing 1400 cm³ of N_2 at atmospheric pressure 101 kPa (1 Pa = 1 N/m²) and room temperature 295 K. How many molecules are in the bottle? What is the mass of the gas?

Mixtures of gases

Since the gas molecules in the ideal gas model do not interact (assumption 4), we can consider gases that contain a mixture of chemical species as being completely separate gases that happen to occupy the same volume. The ideal gas law applies to each species separately

$$P_s V = N_s k_B T \tag{T2.14}$$

where T is the common temperature of the mixture, N_s is the number of molecules of species s in the common volume V, and P_s is the **partial pressure** that the molecules of that species s exert on the container walls. Note that we can calculate N_s for each species by using equation T2.13 and the molecular mass $M_{A,s}$ for that species. Just as the total number of molecules in the mixture is the sum of N_s over all species, the total pressure exerted by the mixture is just the sum of the partial pressures:

$$P = \sum_{\text{all species}} P_s \tag{T2.15}$$

T2.2   Temperature and Energy

Temperature is linked to molecular kinetic energy

Equation T2.6 has another very important implication. Because the gas molecules move in a totally random fashion (assumption 6), if we had chosen our coordinate axes so that the piston face were perpendicular to the y direction instead of to the x direction, we would have found that $[mv_y^2]_{avg} = k_B T$ instead of $[mv_x^2]_{avg} = k_B T$. The same argument applies to the z direction, so we must have

$$[mv_x^2]_{avg} = [mv_y^2]_{avg} = [mv_z^2]_{avg} = k_B T \tag{T2.16}$$

Now, if we again assume that the ith molecule in our gas of N molecules has mass m_i and velocity components $v_{i,x}$, $v_{i,y}$, and $v_{i,z}$, then the definition of the average means that

$$[mv_x^2]_{avg} + [mv_y^2]_{avg} + [mv_z^2]_{avg}$$

$$= \frac{1}{N}\sum_{i=1}^{N} m_i v_{i,x}^2 + \frac{1}{N}\sum_{i=1}^{N} m_i v_{i,y}^2 + \frac{1}{N}\sum_{i=1}^{N} m_i v_{i,z}^2$$

$$= \frac{1}{N}\sum_{i=1}^{N} \left(m_i v_{i,x}^2 + m_i v_{i,y}^2 + m_i v_{i,z}^2 \right) \tag{T2.17}$$

since it does not matter how we group the terms in the sum. Since $m_i v_{i,x}^2 + m_i v_{i,y}^2 + m_i v_{i,z}^2 = m_i(v_{i,x}^2 + v_{i,y}^2 + v_{i,z}^2) = m_i v_i^2$, equation T2.16 implies that

$$K_{avg} = \frac{1}{2}[mv^2]_{avg} = \frac{3}{2}k_B T \tag{T2.18}$$

Purpose: This equation links the average molecular kinetic energy K_{avg} in an ideal gas to the gas's absolute temperature T.
Symbols: k_B is Boltzmann's constant.
Limitations: This equation applies when the ideal gas model is a good approximation.

Exercise T2X.3

Fill in the minor missing steps leading to equation T2.18.

Equation T2.18 is an extremely important equation. It tells us that there is a direct and simple link between the temperature of an ideal gas and the kinetic energy of its molecules: The Boltzmann constant k_B merely specifies the connection between our arbitrarily chosen temperature and energy units! If we had a useful way of measuring molecular kinetic energies directly, in fact we could have *defined* temperature in terms of the average molecular kinetic energy of an ideal gas. Astrophysicists often express the temperatures of astronomical objects in terms of the value of $k_B T$ in electronvolts (abbreviated as eV) rather than T in kelvins, because the value of $k_B T$ is directly connected to the typical energies of molecules in the objects and therefore to their physical behavior.

Exercise T2X.4

(a) Show that the value of $k_B T$ at room temperature (295 K) is $0.0254\,\text{eV} \approx 1/40$ eV (this is worth memorizing). (b) What is the value of $k_B T$ on the surface of the sun?

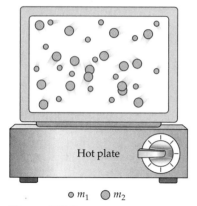

Figure T2.2
A container holding molecules of two different masses m_1 and m_2 sits on a hot plate that we can use to control its temperature.

Equation T2.18 also clarifies the physical meaning of *absolute zero*. If temperature is directly proportional to the average kinetic energy of a gas molecule, then absolute zero in a gas implies that all the gas molecules all have exactly zero kinetic energy (i.e., they are exactly at rest). Clearly, no lower temperature is possible (K_{avg} cannot be negative!).

T2.3 Molecular Speeds and Brownian Motion

Equation T2.18 applies to mixtures of gases as well as pure gases, since we did not anywhere assume that all the molecules in the container had the same mass. For example, imagine a mixture of two gases, with molecules of masses m_1 and m_2, respectively (see figure T2.2). Equation T2.18 implies that if the two gases are in thermal equilibrium with the walls and each other, the average molecular kinetic energies must be the same for both gases:

$$\frac{1}{2}m_1 \left[v_1^2\right]_{\text{avg}} = \frac{3}{2}k_B T = \frac{1}{2}m_2 \left[v_2^2\right]_{\text{avg}} \tag{T2.19}$$

This means that the more massive a molecule is, the smaller its average squared speed will be. The small molecules in figure T2.2 will thus zip around their container quite rapidly, while the big molecules will amble about at lower speeds.

An approximate measure of the average speed of a molecule is the square root of its average squared speed. We can easily calculate this for a molecule with a given mass m, using equation T2.18: Since $\frac{1}{2}m[v^2]_{\text{avg}} = (3/2)k_B T$,

The rms speed of a gas molecule

$$v_{\text{rms}} \equiv \sqrt{[v^2]_{\text{avg}}} = \sqrt{\frac{3k_B T}{m}} \tag{T2.20}$$

Purpose: This equation specifies the **root-mean-square, or rms, speed** v_{rms} of a gas molecule with mass m in a gas with absolute temperature T.
Symbols: k_B is Boltzmann's constant.
Limitations: This equation strictly applies only to an ideal gas.

The rms speed (which, you've got to admit, is easier to say than "the square root of the average of the squared molecular speed") is technically not exactly the same as the molecule's *average speed* but is of the same order of magnitude (we will learn more about this in chapter T7). The rms speed for helium atoms at room temperature (295 K) turns out to be about 1360 m/s, whereas the rms speed for nitrogen molecules is about 510 m/s. Direct measurements of molecular speeds in gases in the ideal gas limit confirm the validity of equations T2.18 and T2.20.

Note that the speed of sound (at 340 m/s) is about the same order of magnitude of v_{rms} for nitrogen. This makes sense: since air molecules are not physically in contact with one another most of the time and do not influence one another at a distance, the only way that information about a disturbance in the air can move from one point to another nearby is if that information is carried by molecules that physically travel between those points.

Exercise T2X.5

Verify that v_{rms} is about 1360 m/s for helium and 510 m/s for N_2 molecules at 295 K. (Helium has a molar mass of 4 g/mol and N_2 a mass of 28 g/mol.)

The equality of average kinetic energies applies even when the ratio of m_2 to m_1 is very large—in fact, even in the case where the "molecules" are actually tiny but visible particles of matter. In 1827, the English botanist Robert Brown discovered that grains of pollen suspended in water seemed to move continuously and randomly. At first, this motion (which came to be called **brownian motion**) were taken as a sign of life, but careful experiments showed that small inorganic particles behaved similarly.

In 1905, Albert Einstein (in the same issue of *Annalen der Physik* that contained his first paper on special relativity) published a detailed quantitative description of brownian motion based on equation T2.17. Einstein argued that the motion that Brown had observed was due to collisions between water molecules and the pollen grains, and that the pollen grains were exhibiting the random motion to be expected from equation T2.20.

Moreover, Einstein noted, it should be possible to determine Avogadro's number by studying brownian motion. If molecules are extremely small (which is what happens if Avogadro's number is large), then many molecules will strike the grain from many directions at once during a given time interval, and the forces that the molecules exert will mostly cancel out. On the other hand, if molecules are few in number and large in mass, the particles striking the grain at a given time more likely will exert unbalanced forces, leading to greater jostling of the grain. Einstein derived a quantitative relation between a grain's mass, its average displacement during jostling, and Avogadro's number, making it possible to deduce Avogadro's number from measurements of brownian motion.

In 1908, Jean Perrin published results from a brilliant series of experiments using this and other aspects of brownian motion to compute Avogadro's number, arriving at a figure of 6×10^{23} molecules per mole. His work, which earned him the 1926 Nobel Prize in physics, convinced the skeptics in the physics community that atoms were real objects, not simply mathematical conveniences. These results also proved wrong the critics of Boltzmann's work in statistical mechanics (sadly, they were published two years after Boltzmann's death). Figure T2.3 shows one of Perrin's sketches of a suspended particle's motion.

Brownian motion

Einstein's method for finding Avogadro's number

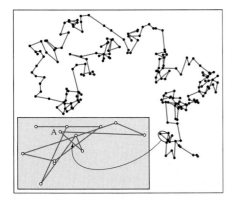

Figure T2.3
A sketch by Jean Perrin showing the position of a particle suspended in water every 30 s. The inset at the bottom shows its positions every 3 s as it travels between two points in the larger diagram.

T2.4   The Thermal Energy of a Gas

The total thermal energy of a gas in our simple point-particle model of a gas is nothing more than the sum of those particles' kinetic energies of random motion. Equation T2.18 implies that this thermal energy is directly proportional to the temperature T of the gas and depends *only* on N and T:

The thermal energy of a point-particle gas

Real gases are more complicated

$$U = \sum_{i=1}^{N} K_i = N\left(\frac{1}{N}\sum_{i=1}^{N} K_i\right) = NK_{\text{avg}} = \frac{3}{2}Nk_B T \qquad \text{(T2.21)}$$

Equation T2.21 makes a testable prediction about the thermal energy of a gas that we might not have suspected without having gone through this argument. If we put a tiny amount of energy dU into a gas, its temperature will increase by dT such that

$$dU = \frac{3}{2}Nk_B\,dT \qquad \Rightarrow \qquad \frac{dU}{dT} = \frac{3}{2}Nk_B \qquad \text{(T2.22)}$$

We can test this by putting a known amount of gas in a fixed volume, adding a small, known amount of energy dU to the gas, and measuring the resulting change in temperature dT. Do real gases behave as predicted?

The answer is yes and no. Table T2.1 shows measured values of dU/dT expressed as a multiple of Nk_B for various kinds of gases (evaluated at 300 K, except in the case of water). Note that **monatomic** gases (such as helium, argon, and neon) whose molecules are comprised of a single atom do seem to have values reasonably consistent with our prediction. On the other hand, **diatomic** gases (whose molecules consist of exactly two atoms) seem to have

Table T2.1 dU/dT for various kinds of gases

Monatomic	dU/dT
Helium	$1.51Nk_B$
Neon	$1.53Nk_B$
Argon	$1.50Nk_B$
Krypton	$1.48Nk_B$
Xenon	$1.51Nk_B$

Diatomic	dU/dT
Hydrogen (H_2)	$2.46Nk_B$
Nitrogen (N_2)	$2.50Nk_B$
Oxygen (O_2)	$2.52Nk_B$
Carbon monoxide (CO)	$2.49Nk_B$

Polyatomic	dU/dT
Water (H_2O)	$3.25Nk_B$
Methane (CH_4)	$3.26Nk_B$
Carbon dioxide (CO_2)	$3.43Nk_B$
Hydrogen sulfide (H_2S)	$3.29Nk_B$

Adapted from Serway and Beichner, *Physics*, 5th ed., p. 647, and other sources.

a value for dU/dT that is within a few percent of $(5/2)Nk_B$. **Polyatomic** gases (whose molecules are composed of more than two atoms) have a broader range of values above $3Nk_B$.

Direct measurements of molecular velocities are consistent with $K_{avg} = (3/2)k_BT$ for all types of gases in the ideal gas limit. But equations T2.21 and T2.22 are based on an *additional* assumption that the total thermal energy of a gas is simply the sum of the molecular kinetic energies. The results in table T2.1 show that while this assumption seems pretty good for monatomic gases, it is clear that we have to put more energy into a diatomic or polyatomic gas to change its temperature by a given amount than can be accounted for by the change in its total molecular kinetic energy.

One plausible model is that diatomic and polyatomic molecules can store additional thermal energy in the form of rotational and/or vibrational energy. Qualitatively, collisions between such molecules and a container wall can convert some of the molecule's center-of-mass kinetic energy to rotational kinetic energy or vibrational energy (or vice versa), and this allows these energy storage modes to store an average energy comparable to the average amount of kinetic energy available in a collision. So, if we put energy into a diatomic or polyatomic gas, only some of the energy goes to increase the average molecular kinetic energy K_{avg}, and the rest is channeled to rotational and/or vibrational energy. Since $T = (2/3)K_{avg}/k_B$ for all gases, this means that the temperature change dT for a diatomic or polyatomic gas is smaller than it would be for a monatomic gas absorbing the same amount of energy dU.

As we will see in chapter T7, this model is basically correct. If we plot dU/dT for a diatomic gas such as hydrogen as a function of temperature over a large temperature range, though, we see that dU/dT behaves in a very curious manner that this model does not explain (see figure T2.4). Between hydrogen's boiling point at 20 K and about 80 K, $dU/dT \approx (3/2)Nk_B$, as if our hypothetical rotational and vibrational energy storage modes were completely switched off! As temperature increases, we see that dU/dT increases (with broad transition regions) in steps of Nk_B to $(5/2)Nk_B$ and then toward $(7/2)Nk_B$. [While hydrogen disassociates at 2000 K at normal pressure, if one measures dU/dT at higher temperatures and pressures, it does eventually level off at $(7/2)Nk_B$.] It is as if we are switching on the hydrogen molecule's extra rotational and/or vibrational energy storage modes one at a time as T

Explaining the results

A mystery

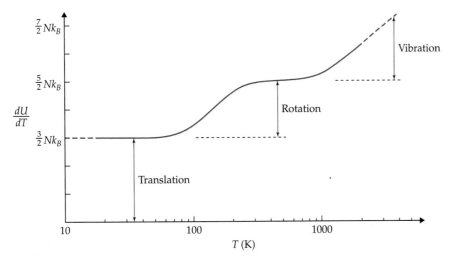

Figure T2.4
The heat capacity of hydrogen gas at standard pressure as a function of temperature. (Adapted from Schroeder, *Thermal Physics*, Addison-Wesley, 2000, p. 30.)

increases, and each additional mode stores an average energy of $k_B T$ per molecule once it is fully "on." In the transitional regions, we can think of the mode as being partially on: it stores *some* energy but not as much as it might.

But if a molecule can rotate or vibrate at all, our simple model suggests that it would do so at *all* temperatures. Why are these modes turned on only if the temperature is sufficiently high? We will solve this mystery using statistical methods later in the unit.

We see empirically, though, that (except for transitional temperatures) a gas's total thermal energy U in the ideal gas limit appears to be an integer multiple of $\frac{1}{2} N k_B T$. In general, we can say that

$$U = \frac{f}{2} N k_B T \qquad \text{(T2.23)}$$

Purpose: This equation specifies the total thermal energy U of a sample of ideal gas containing N molecules at absolute temperature T.

Symbols: k_B is Boltzmann's constant. Near room temperature, $f \approx 3$ for purely monatomic gases, $f \approx 5$ for purely diatomic gases, and $f > 6$ for polyatomic gases. (Physicists refer to f the number of molecular *degrees of freedom* for reasons beyond our scope here.)

Limitations: Equation T2.23 only applies to real gases in the ideal gas limit. The variable f is never exactly an integer, but it is very close to being an integer for monatomic and diatomic gases in certain temperature ranges.

We will find this expression very useful in chapters T3, T7, and T8.

Exercise T2X.6

A liter (abbreviated as L) [$= 10^{-3}$ m$^3 \approx 1$ quart (qt)] of gas at standard pressure contains roughly 2.48×10^{22} molecules at room temperature (295 K). Roughly how much thermal energy is stored in such a volume of air? About how fast would a full soda can (whose mass is about 0.34 kg) have to travel to have this much kinetic energy?

T2.5   Solids and Liquids

No easy models for liquids

We have seen that it is easy to build a simple but pretty useful newtonian model of a gas. This is *not* the case for liquids. The main difference between liquids and gases is that the molecules in a liquid, while somewhat free to move, are also close enough to touch, making the liquid comparatively incompressible. The fact that the molecules in a liquid are almost always in contact means that several of the assumptions we used in "deriving" the ideal gas law (specifically, the assumptions that each molecule can be treated in isolation and does not significantly interact) do not even approximately apply. The significant but complex interactions between molecules make it hard to construct a microscopic model of liquids having the quantitative accuracy of the ideal gas model. Still, the most general principles that apply to gases apply to liquids as well: empirically liquids do store thermal energy (on the order of a few times $k_B T$ per molecule) in the form of kinetic energy of molecular motion and the potential energy of intermolecular interactions. The brownian motion experiments also indirectly support the idea that the average center-of-mass *kinetic* energy of a water molecule is $(3/2) N k_B T$, just as it is for gases.

In the case of solids, these intermolecular interactions become so strong that the molecules are essentially locked into fixed locations in a lattice. The molecules in such a solid store thermal energy in the form of vibrational kinetic and potential energy. As in liquids and gases, the average thermal energy stored per molecule in a solid increases fairly uniformly with increasing temperature. Indeed, for monatomic solids at reasonably high temperatures, the measured value of dU/dT has a nearly constant value of almost exactly $3Nk_B$ (see table T2.2), a result called the **law of Dulong and Petit**. If this is reasonably true for all temperatures and we assume that $U = 0$ at $T = 0$, this means that the total thermal energy in a monatomic solid is given by

$$dU \approx 3Nk_B\, dT \quad \Rightarrow \quad U \approx 3Nk_B \int_0^T dT = 3Nk_BT \qquad \text{(T2.24)}$$

which is almost exactly twice the energy stored in a monatomic gas.

That the values of dU/dT shown in table T2.2 should be so close to the same integer multiple of Nk_B tantalizingly suggests that there might be a simple model for monatomic solids analogous to the model of monatomic gases we have developed in this chapter. We will, in fact, develop such a model in the next few chapters.

T2.6   Conclusions

The ideal gas model presented in this chapter helps us see that there is a direct link between the constant-volume gas temperature T defined in chapter T1 and the average microscopic kinetic energy of gas molecules. Qualitatively at least, the same idea applies to liquids and solids as well: *Temperature is connected with molecular motion.*

However, making more than qualitative predictions about this connection in a given physical situation requires having a good model for that situation. The ideal gas model developed in this chapter successfully predicts that sufficiently rarefied gases will satisfy the ideal gas law and that their average molecular kinetic energy is $K_{avg} = (3/2)k_BT$. However, the empirical results presented in the last few sections raise a number of questions we have yet to answer:

1. Why do the extra energy storage modes available to diatomic and polyatomic gas molecules only switch on at sufficiently high temperatures?
2. For a diatomic gas, why do these modes seem to store an energy per molecule of almost exactly k_BT when they are fully switched on?
3. Why is the average energy per atom in a monatomic solid so close to $3k_BT$ at normal temperatures?

Our gas model has also not yet provided any help in understanding the core mystery of this unit:

4. Why are some energy transfers irreversible?

It turns out that the straightest path to solving all these mysteries involves working with a quantum-mechanical model for the monatomic solid. By closely examining how this model behaves, we will be able to develop the conceptual and mathematical tools needed to answer all these questions. This will be our task, starting with chapter T4.

However, we can learn still more from the ideal gas model, things that will provide a useful background for what is to come. In chapter T3, we will use our ideal model to study processes in which the properties of a sample of gas change with time.

Solids

Table T2.2   Measured values of dU/dT for some monatomic solids

Solid	dU/dT
Lead	$3.18Nk_B$
Gold	$3.06Nk_B$
Silver	$3.00Nk_B$
Copper	$2.95Nk_B$
Iron	$3.01Nk_B$
Aluminum	$2.92Nk_B$

Adapted from Serway, *Physics*, 3d ed., p. 530, and other sources.

TWO-MINUTE PROBLEMS

T2T.1 If the speed of a molecule in a container doubles (other things remaining the same), the average pressure that this molecule exerts on the container wall
A. Remains the same
B. Doubles
C. Quadruples
D. It is impossible to say

T2T.2 The ideal gas model assumes that molecules have infinitesimal size. Consider one molecule bouncing around in the container. As the molecule's size increases relative to the size of the container, how will the average pressure P that it exerts change, other things being held constant?
A. P increases.
B. P does not change.
C. P decreases.
D. It is impossible to say.

T2T.3 The average x-velocity of molecules in a container of gas at rest is zero (T or F). The average molecular speed is zero (T or F).

T2T.4 $[v_x]_{avg} = \sqrt{[v_x^2]_{avg}}$ (T or F).

T2T.5 Two identical rooms, A and B, are sealed except for an open doorway between them. Room A is warmer than B. Which room has the greater number of molecules?
A. Room A
B. Room B
C. Both have the same number of molecules.

T2T.6 One mole of hydrogen gas and $\frac{1}{2}$ mol of helium gas are mixed in a container and maintained at a fixed temperature T. How does the total pressure P_H exerted by the hydrogen molecules compare to the total pressure P_{He} exerted by the helium molecules? Note that a molecule (atom) of helium has twice the mass of an H_2 molecule.
A. $P_H = 2P_{He}$
B. $P_H = \sqrt{2}P_{He}$
C. $P_H = P_{He}$
D. $P_H = \sqrt{\frac{1}{2}}P_{He}$
E. $P_H = \frac{1}{2}P_{He}$
F. Other (specify)

T2T.7 N molecules of a monatomic gas A is mixed with the same number of molecules of a diatomic gas B in a container, and the mixture is held at a constant temperature T (which is about equal to room temperature). The molecules of gas A have mass m_A, and the molecules of gas B have mass $m_B > m_A$. Which gas has (1) the greater molecular v_{rms}, (2) the greater average kinetic energy per molecule, (3) the greater total thermal energy U, and (4) the greater pressure P? In each case, choose from one of the four answers below.
A. The quantity is greater for gas A.
B. The quantity is greater for gas B.
C. The quantity is the same for both gases.
D. It is impossible to compare the quantities.

T2T.8 Imagine that when we add 20.8 J to 1 mol of a certain gas, its temperature is observed to increase by 1 K. Which of the following statements is true?
A. This gas is monatomic.
B. This gas is diatomic.
C. This gas is polyatomic.
D. We cannot determine the type of gas without more information.
E. The results described are impossible.

HOMEWORK PROBLEMS

Basic Skills

T2B.1 In interstellar space, there is about one H_2 molecule per cubic centimeter. The temperature of deep space is about 3 K. What is the pressure of this interstellar gas? How does this compare to the pressure of the best vacuum that can presently be attained in the laboratory ($\approx 10^{-13}$ Pa)?

T2B.2 Estimate the number of air molecules in your dorm room or bedroom. (You will have to estimate the room's size.) Does it matter what kinds of molecules the air in your room actually contains? Why or why not?

T2B.3 Dry air consists of 78% nitrogen, 21% oxygen, and 1% argon by weight. Show that the mass of 1 mol of air is about 29.0 g.

T2B.4 Consider a gas whose molecules have an average kinetic energy of 1 eV. What is the temperature of the gas?

T2B.5 Find the rms speed of CO_2 molecules in a room at room temperature (295 K). (Carbon atoms have 12 and oxygen atoms have 16 nucleons in their nuclei.)

T2B.6 The rms speed of Cl_2 molecules in a sample of gas is about 180 m/s. What is the temperature of the gas? (Chlorine atoms have 35 nucleons in their nuclei.)

T2B.7 What is the total thermal energy of 0.40 g of helium gas inside a balloon at room temperature? How fast would you have to run to have that much kinetic energy?

T2B.8 What is the approximate total thermal energy in a 19.7-g nugget of gold at room temperature? (Gold has an atomic weight of 197 g/mol.)

Synthetic

T2S.1 (a) Show how the law of partial pressures given in equation T2.14 follows mathematically from equation T2.13.
(b) Use the ideal gas model to explain qualitatively why this *physically* makes sense.

T2S.2 *Boyle's law* (first stated by Robert Boyle in 1662) claims that the pressure P of a given sample of gas at a given temperature is inversely proportional to its volume V. *Gay-Lussac's law* (first stated by Joseph-Louis Gay-Lussac in 1802) claims that the volume V of a gas is proportional to its absolute temperature T at a fixed pressure. Amedeo Avogadro argued in 1811 that at constant pressure and temperature, the volume occupied by a sample of gas is proportional to the number of molecules it contains, independent of the type of gas. Show how all three of these laws follow from the ideal gas law.

T2S.3 (a) The average velocity of a molecule in a gas at rest is zero. How do we know this?

(b) The rms speed of a molecule in a gas at rest is not zero. How is this possible?

T2S.4 Imagine that during a hailstorm, hailstones with an average mass of 1 g and an average vertical velocity of 10 m/s hit the roof of your car at a rate of 100 hits per

second per square meter; and assume that they bounce elastically off the roof. What is the approximate pressure that these hailstones exert on your roof? How does this compare to air pressure? Estimate the total force exerted by the hail on your roof.

T2S.5 (a) The upward buoyant force on a balloon is equal to the weight of the air that it displaces. Assume that the mass of the unfilled balloon and its payload is M. Argue that if the buoyant force is approximately in balance with the gravitational force on the balloon (so that the balloon floats), we must have

$$\rho_o - \rho_i = \frac{M}{V} \qquad (T2.25)$$

where ρ_o is the density of the air outside the balloon, ρ_i is the density of the gas inside the balloon, and V is the volume of the balloon.

(b) Argue that the density of any gas is

$$\rho = \frac{M_A P}{RT} \qquad (T2.26)$$

where M_A is the mass of 1 mol of the gas, P is its pressure, T is its absolute temperature, and $R = N_A k_B = 8.31$ J/K.

(c) In a typical hot-air balloon (see figure T2.5), the bottom of the balloon is open to the surrounding atmosphere, so the pressure of the hot air inside the balloon must be essentially equal to that of the cold air surrounding the balloon. If the air inside a typical hot-air balloon is roughly at $100°C = 373$ K, and the surrounding air is at about $17°C = 290$ K, about how much volume does the balloon have to have? (*Hint:* See problem T2B.3.) What is the radius of a sphere with the same volume? Does your answer make sense, considering the picture?

Figure T2.5
Some hot-air balloons.

Figure T2.6
The International Space Station.

T2S.6 The temperature on Mercury's sunny side is about 700 K, and Mercury's escape speed is about 4300 m/s. Explain why the atmosphere of Mercury, if it has any, cannot contain any hydrogen or helium.

T2S.7 Imagine that we add 250 J of thermal energy to 1 liter (L) (1000 cm^3) of air initially at room temperature and normal pressure. By about how much does its pressure increase if its volume is held constant?

Rich-Context

T2R.1 Edward Purcell, a physicist at the Massachusetts Institute of Technology proposed the following problem as a way to learn about the magnitude of Avogadro's number. A __ of water contains about as many molecules as there are __s of water in all the oceans of the earth. What single word best fits the two blank spaces: *drop, teaspoon, tablespoon, cup, quart, barrel,* or *ton?* Oceans cover about 70% of the earth's surface area and have an average depth of about 5 km.

T2R.2 One of the hazards of being in space is the possibility that a meteoroid might punch a hole in the vehicle you are traveling in, allowing the air inside to leak away into the vacuum of space. Roughly how long would you have to live if a meteoroid punched a hole with an area $a = 1$ cm^2 in your spacecraft? The answer to this question may be interesting not only to you but also to the *designers* of the spacecraft. For example, providing hole-patching materials is pointless if the astronauts will only live for a few seconds.

To make the problem more concrete, assume we are talking about a reasonably sized module of the In-

ternational Space Station (see figure T2.6). Assume that this module is sealed off from the rest of the other modules; it is shaped like a cylinder roughly 4 m in diameter and 10 m long; and the hole has been punched in one of the end faces of the cylinder.

(a) Roughly how often does a given molecule hit the end face of the module? What is the approximate probability that if it does, it will go through the hole? Combine the answers to these questions to argue that the number of molecules of air that escape through the hole per unit time is given by

$$\left| \frac{dN}{dt} \right| \approx \frac{a \, |v_x|_{\text{avg}}}{2V} N \qquad (T2.27)$$

where V is the volume of the module, a is the area of the hole, N is the number of molecules in the module, v_x is the x component of a given molecule's velocity, and the average is taken over all molecules. Note that it is the average of the absolute value of v_x, not the average of v_x itself (why not?).

(b) Show that this equation implies that $N(t) = N_0 e^{-t/\tau}$, where N_0 is the number of molecules present at time $t = 0$ and $\tau \equiv 2V/a\,|v_x|_{\text{avg}}$.

(c) Given this, how long do you think someone can survive before blacking out (in terms of τ)? Explain how you are making your estimate.

(d) It is hard to calculate $|v_x|_{\text{avg}}$ exactly, but argue that $|v_x|_{\text{avg}} \approx \sqrt{k_B T/m}$, where m is the mass of a molecule. Why is this an approximation?

(e) Put these results together to answer the question.

Advanced

T2A.1 (a) Consider a cylindrical parcel of air of area A and infinitesimal height dz. If this air parcel is to remain stationary, the difference between the total pressure forces exerted on its top and bottom faces must be equal to its weight. Use this information and the ideal gas law to show that

$$\frac{dP}{dz} = -\frac{mg}{k_B T} P \qquad (T2.28)$$

where m is the mass of an air molecule, g is the gravitational field strength, k_B is Boltzmann's constant, T is the absolute temperature, and P is the air pressure at the vertical position z corresponding to the center of the parcel.

(b) Assuming that T is independent of height in the earth's atmosphere (which is a pretty crude approximation), find a formula for the air pressure P as a function of altitude z. Use this to estimate the air pressure at the top of Mount Everest at $z = 8840$ m above sea level as a fraction of the pressure at sea level.

ANSWERS TO EXERCISES

T2X.1 If the molecule's x-velocity increases by a factor of 2, the *change* in its x-velocity when it hits the piston also doubles, so the impulse delivered to the piston doubles. But the time between collisions also *decreases* by a factor of 2, since it takes one-half the time for the molecule to cover the round trip at twice the speed. So the molecule delivers twice the impulse twice as often, implying that the average force and thus the pressure increase by a factor of 4.

T2X.2 According to the ideal gas law, the number of molecules is

$$N = \frac{PV}{k_B T} = \frac{(101,000\,\text{N/m}^2)(1400\,\text{cm}^3)}{(1.38 \times 10^{-23}\,\text{J/K})(295\,\text{K})}\left(\frac{1\,\text{m}}{100\,\text{cm}}\right)^3$$

$$\times \left(\frac{1\,\text{J}}{1\,\text{N m}}\right) = 3.47 \times 10^{22} \qquad (T2.29)$$

Since the mass of Avogadro's number of nitrogen molecules is $M_A = 28$ g/mol, the mass of our sample, according to equation T2.13, is

$$M = \frac{M_A}{N_A} N = \frac{28\,\text{g/mol}}{6.02 \times 10^{23}\,/\text{mol}} 3.47 \times 10^{22}$$

$$= 1.62\,\text{g} \qquad (T2.30)$$

T2X.3 Note that

$$K_{\text{avg}} = \frac{1}{2}[mv^2]_{\text{avg}} \equiv \frac{1}{2}\left(\frac{1}{N}\sum_{i=1}^{N} m_i v_i^2\right)$$

$$= \frac{1}{2}\left([mv_x^2]_{\text{avg}} + [mv_y^2]_{\text{avg}} + [mv_z^2]_{\text{avg}}\right) \qquad (T2.31)$$

according to equation T2.17. But equation T2.16 implies that last line is equal to $3k_B T$. Therefore, $K_{\text{avg}} = (3/2)k_B T$.

T2X.4 At room temperature

$$k_B T = (1.38 \times 10^{-23}\,\text{J/K})(295\,\text{K})\left(\frac{1\,\text{eV}}{1.60. \times 10^{-19}\,\text{J}}\right)$$

$$= 0.0254\,\text{eV} \qquad (T2.32)$$

According to table T1.1, the temperature at the surface of the sun is 5800 K, so

$$k_B T = (1.38 \times 10^{-23}\,\text{J/K})(5800\,\text{K})\left(\frac{1\,\text{eV}}{1.60 \times 10^{-19}\,\text{J}}\right)$$

$$= 0.50\,\text{eV} \qquad (T2.33)$$

a nice round number!

T2X.5 According to equation T2.20, the root-mean-square speed of helium (which has a molar mass of 4 g/mol) at $T = 295$ K is

$$v_{\text{rms}} = \sqrt{\frac{3k_B T}{m}} = \sqrt{\frac{3k_B T}{M_A/N_A}} = \sqrt{\frac{3RT}{M_A}}$$

$$= \sqrt{\frac{3(8.31\,\text{J K}^{-1}\,\text{mol}^{-1})(295\,\text{K})}{0.004\,\text{kg/mol}}\left(\frac{1\,\text{kg m}^2/\text{s}^2}{1\,\text{J}}\right)}$$

$$= 1360\,\text{m/s} \qquad (T2.34)$$

where m is the mass of a molecule, N_A is Avogadro's number, $M_A = mN_A$ is the molar mass of the molecule, and $R = N_A k_B$ is the gas constant (see equation T2.11). Since nitrogen has a molar mass of 28 g/mol, which is 7 times larger than that of helium, the rms speed will be $\sqrt{7}$ times smaller, or about 510 m/s.

T2X.6 Since air consists almost entirely of diatomic molecules, the total thermal energy stored by 2.48×10^{22} molecules at 295 K is

$$U = \frac{5}{2}Nk_B T$$

$$= \frac{5}{2}(2.48 \times 10^{22})(1.38 \times 10^{-23}\,\text{J/K})(295\,\text{K}) \approx 250\,\text{J} \qquad (T2.35)$$

To have this much kinetic energy, a soda can with mass $m = 0.34$ kg would have to be traveling at a speed v such that

$$U = K = \frac{1}{2}mv^2 \quad \Rightarrow \quad v = \sqrt{\frac{2U}{m}}$$

$$= \sqrt{\frac{2(250\,\text{J})}{0.34\,\text{kg}}\left(\frac{1\,\text{kg m}^2/\text{s}^2}{1\,\text{J}}\right)}$$

$$= 38.5\,\text{m/s} \qquad (T2.36)$$

This is about 86 mi/h! (I would not want to stand in the way of this soda can.)

T3

Gas Processes

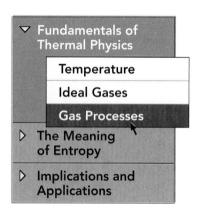

Chapter Overview

Introduction

In chapter T2, we developed a simple newtonian model for an ideal gas. In this chapter we will use the model to explore thermal *processes* in which the properties of a gas change with time. This chapter provides essential background for the rest of the unit, especially for chapters T8 and T9.

Section T3.1:   Review of Heat and Work

The technical terms **heat** and **work** both describe energy flowing across a system boundary during a process. The *heat Q* is that part of the energy flow driven by a temperature difference between the system and its surroundings; the *work W* refers to any other energy flow across the boundary (note that this is *not* the same as the "k-work" [dk]defined in unit C). The energy stored *inside* a system boundary is the system's **thermal energy** U. These quantities are linked by the **first law of thermodynamics:**

$$\Delta U = Q + W \qquad\qquad\text{(T3.1)}$$

Purpose: This equation expresses the law of conservation of energy in the context of thermodynamic systems.

Symbols: ΔU is the change in a system's thermal energy in a given process; Q and W are the heat and work, respectively, that have flowed into or out of the system during that process.

Limitations: There are none. Heat and work are mutually exclusive, but together they include all the ways that energy can flow into or out of a system.

Note: In this text, both Q and W are positive when energy flows *into* the system. Some other texts adopt a different sign convention for W.

Section T3.2:   Work During Expansion or Compression

Because a gas molecule will rebound from a moving piston with a different energy than it had originally, changing a gas's volume involves work, which we can calculate as follows:

$$dW = -P\,dV \qquad\qquad\text{(T3.6)}$$

Purpose: This equation expresses the thermodynamic work dW done on a gas during an infinitesimal volume change dV.

Symbols: P is the gas pressure during the volume change.

Limitations: The volume change must be small enough that P does not change significantly during the process, and slow enough that the forces on the piston are essentially in balance, the piston's kinetic energy is negligible, and the gas is in equilibrium with itself.

Note: To find the work W involved during a process in which the pressure varies, one must integrate this expression; see section T3.5.

Section T3.3:   The State of a Gas

A gas's **macroscopic properties** (such as its pressure P, volume V, temperature T, number of molecules N, and thermal energy U) are quantities that we can measure macroscopically (i.e., without detailed measurements of what its molecules are doing). It turns out that knowing a suitably chosen *triplet* of such properties (such as P, V, and N) allows us to calculate a gas's other macroscopic properties. Such a triplet describes the gas's **macrostate.**

Section T3.4:   *P–V* Diagrams and Constrained Processes

So, when N is known and fixed, knowing P and V determines a gas's macrostate. We can represent such a state by a point on a graph of P versus V (a **P–V diagram**). During a **quasistatic gas process** (a process slow enough that even as P changes in time, its value remains uniform throughout the gas), the point representing the gas's state traces out a curve on a $P–V$ diagram.

A **constrained process** is a process during which we constrain the gas in a certain way. During an **isothermal process,** we constrain the gas's temperature to be constant; during an **isobaric** process we hold its pressure constant; during an **isochoric process,** we keep its volume constant; during an **adiabatic process** we keep heat energy from flowing into or out of the gas (see figure T3.5). These are useful approximations for realistic gas processes.

Section T3.5:   Computing the Work

To compute the work W done during a process during which P changes significantly, we have to evaluate the integral

$$W = -\int P\,dV \qquad\qquad (T3.8)$$

We can evaluate this integral mathematically if we can express P as a function of V, which we can do easily for isochoric, isobaric, and isothermal processes (see equations T3.9, T3.10, and T3.12). But the most important thing to note is that the magnitude of this integral corresponds to the *area under the process's curve on a P–V diagram,* an idea we can use to quickly estimate and compare the work involved in various processes.

Section T3.6:   Adiabatic Processes

Determining how P, V, and T are related in an adiabatic process is trickier than for the other processes. Using the newtonian gas model and some simple arguments about how molecules are affected by a moving piston, however, one can show that for an adiabatic process

$$TV^{\gamma-1} = \text{constant} \qquad\qquad (T3.23a)$$
$$PV^{\gamma} = \text{constant} \qquad\qquad (T3.23b)$$

> **Purpose:**   This equation specifies how a gas's temperature T and pressure P vary with its volume V during an adiabatic process.
>
> **Symbols:** $\gamma = 1 + 2/f$ is the gas's **adiabatic index,** where f is the number of molecular degrees of freedom for the gas.
>
> **Limitations:**   The process must be adiabatic, the gas must be ideal, and the compression must occur very slowly compared to the thermal speed of the gas molecules.
>
> **Note:**   The values of γ for various types of gases are as follows:
>
> $$\text{Monatomic:}\quad \gamma = \frac{5}{3} = 1.67$$
>
> $$\text{Diatomic:}\quad \gamma = \frac{7}{5} = 1.40$$
>
> $$\text{Polyatomic:}\quad 1 < \gamma \le \frac{4}{3} = 1.33$$

This result will be useful to us in chapters T8 and T9.

T3.1   Review of Heat and Work

I introduced the concept of heat and work in unit C, where my main point was to introduce and emphasize the distinctions between *thermal energy*, *heat*, and *temperature*. In this chapter, I will focus more on the distinction between *heat* and *work*. A brief review of these important technical terms is thus in order.

In thermal physics we are often interested less in an object's total thermal energy than in how that energy changes under certain conditions. As discussed in unit C, knowing how that energy changes involves watching the amount of energy that crosses the object's boundaries. The crucial thermodynamic concepts of *heat* and *work* therefore are both defined to describe *energy transfer across a system boundary*.

Definition of *heat*

When a hot object is placed in contact with a cold object, energy spontaneously flows across the boundary between them until both objects come to have the same temperature. As a result, the thermal energy of the hot object decreases, and the thermal energy of the cold object increases. In physics, **heat** is any energy that crosses the boundary between the two objects *because of a temperature difference across the boundary*. Let me emphasize that to be *heat*, the energy in question *must* (1) be flowing across some kind of boundary between systems (2) as a direct result of a temperature difference across that boundary.

Definition of *work*

We define **work** in thermal physics to be any *other* kind of energy flowing across a system boundary. For example, if I stir a cup of water vigorously and it gets warm as a result, I have not "heated" the water, I have done *work* on it: the mechanical energy flows across the boundary of the water not because of a temperature difference but because of my stirring effort.

Note that *heat* and *work* both refer to energy in transit across a boundary. This sharply distinguishes both terms from *thermal energy*, which refers to energy *enclosed by* the system boundary. Both heat and work flows can contribute to changes in the thermal energy. In fact, conservation of energy implies that

The first law of thermodynamics

$$\Delta U = Q + W \qquad (T3.1)$$

Purpose: This equation expresses the law of conservation of energy in the context of thermodynamic systems.

Symbols: ΔU is the change in a system's thermal energy in a given process; Q and W are the heat and work, respectively, that flowed into or out of the system during that process.

Limitations: There are none. Heat and work are mutually exclusive, but together they include all the ways energy can flow into or out of a system.

Note: In this text, both Q and W are positive when energy flows *into* the system. Some other texts adopt a different sign convention for W.

We call this crucial equation the **first law of thermodynamics.** The definitions of heat and work are illustrated in figure T3.1a.

The distinction between Q and W depends on boundaries

The distinction between heat and work is more a matter of human definition and convenience than one of deep physics: at the microscopic level, an energy flow is simply an energy flow. Moreover, whether we call a flow of energy in a given process heat or work can depend on our choice of system boundary. For example, consider the situation shown in figure T3.1b.

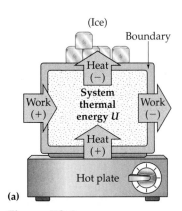
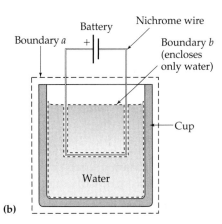

Figure T3.1

(a) Illustration of the definitions of heat and work. (b) Whether an energy flow in a given situation is *heat* or *work* can depend on how we define the system boundary. If we define the system to be enclosed by boundary *a*, the energy flowing into the system is *work*. If we define the system to be enclosed by boundary *b*, the energy flowing into the system is *heat*.

Electricity flows into a piece of nichrome wire that is immersed in a cup of water. The electricity causes the temperature of the wire to increase, which in turn causes the temperature of the water to increase. If we draw the boundary labeled *a* (making "the system" the water and the cup and the resistor), then the energy crossing the boundary is electrical energy and thus *work*. If we draw the boundary labeled *b* (making "the system" the water alone), then the energy crossing the boundary is *heat*, driven by the temperature difference between the wire and the water.

You will become confused by much of what follows if you do not understand the technical distinctions between *heat*, *work*, and *thermal energy*. It is particularly easy to misuse the word *heat* (even some physics texts do!). The following exercise should help develop your ability to distinguish between *heat* and *work*.

Exercise T3X.1

In each process described below, energy flows from object *A* to object *B*. Is the energy flow involved *heat* or *work* (write *Q* or *W* in the blank provided)?

a.   The soup (*B*) in a pan sitting on an electric stove (*A*) gets hot.   _____

b.   Light from an incandescent bulb (*A*) flows to its
     surroundings (*B*).   _____

c.   You (*A*) compress air (*B*) in a bike pump, making it warm.   _____

d.   Your hands (*A*) are warmed when they face a fire (*B*).   _____

e.   The atmosphere (*A*) warms a reentering spacecraft (*A*).   _____

f.   A hot pie (*A*) becomes cooler while sitting in the kitchen (*B*).   _____

g.   Your chair (*B*) becomes warmer after you (*A*) sit in it for
     a while.   _____

h.   A drill bit (*B*) becomes hot after being spun by the drill (*A*).   _____

The difference between work and k-work

Let me remind you that, as discussed in unit C, there is also an important conceptual distinction between thermodynamic work W (which is our focus here) and center-of-mass work or "k-work" $[dK]$ (defined in unit C). This distinction is subtle, because both concepts describe energy transfers. This distinction is also blurred in the conventional literature, where both concepts are called work and given the same symbol W. However, k-work $[dK]$ has a specific and very limited definition: it is the sum of infinitesimal energy transfers $[dK] = \vec{F} \cdot d\vec{r}$ that occur when an interaction exerts a force $\vec{F}$ on the object while the object's center of mass moves an infinitesimal displacement $d\vec{r}$. Thermodynamic work *always* involves an extended object or system, but is otherwise more broadly defined: it is *any* energy flowing across the object's boundary that is not heat, irrespective of whether forces are exerted on the object or whether the object's center of mass moves. Also, k-work (if unopposed by other forces) changes the *kinetic* energy of an object's center of mass, while thermodynamic work (if unopposed by heat energy flowing in the other direction) changes an object's *thermal* energy.

Exercise T3X.2

In each process below, state whether the energy transfer described is k-work or thermodynamic work (write K or W in the space provided).

a. The total kinetic energy of molecules in a bottle of gas increases as the bottle falls in a gravitational field.   _____

b. The total kinetic energy of molecules in a bottle of gas increases as the gas is rapidly stirred by a fan inside the bottle.   _____

c. Air drag on a falling meteorite causes the meteorite's temperature to increase.   _____

d. Air drag on a falling meteorite causes it to slow down (lose kinetic energy).   _____

e. The gas in a bottle gains energy from a spark flowing through it.   _____

f. A small pellet is vaporized by a powerful infrared laser.   _____

T3.2   Work During Expansion or Compression

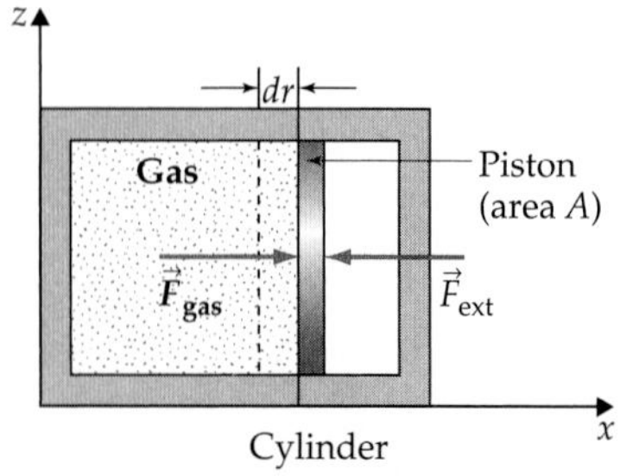

Thermodynamic work can be done on or by a system in a variety of ways, but in much of this unit we will be primarily concerned with work done as a result of the compression or expansion of a gas. Understanding how work is involved in expansion and compression will be especially important to us at the end of the unit, when we discuss how heat can be converted to mechanical energy by using a **heat engine** (such as a gasoline or steam engine). Most heat engines use expansion and compression of some gas to perform this conversion.

Imagine that some external interaction exerts a force $\vec{F}_{ext}$ on a piston that pushes it *slowly* an infinitesimal displacement $d\vec{r}$, compressing a cylinder of gas (see figure T3.2). The energy (k-work) this interaction contributes to the

Figure T3.2

Slowly compressing gas in a cylinder.

piston is

$$[dK] = \vec{F}_{\text{ext}} \cdot d\vec{r} = F_{\text{ext}} \, dr \qquad (T3.2)$$

since $\vec{F}_{\text{ext}}$ and $d\vec{r}$ point in the same direction in this case. Now, the gas in the cylinder exerts an opposing force $\vec{F}_{\text{gas}}$ against the piston. If we push the piston *slowly* enough that its speed remains essentially constant and its kinetic energy change is insignificant, the piston's interaction with the gas must be *extracting* energy from the piston at the same rate as the external interaction supplies it. Where does this energy go? It must ultimately go to thermal energy in the gas. Since this energy flow into the gas is not due to a temperature difference, it is *work*. The point is that in this case the infinitesimal thermodynamic work done on the gas during such an infinitesimal slow compression is equal to the k-work done by the external force on the piston

$$dW = F_{\text{ext}} \, dr \qquad (T3.3)$$

In a compression that is very slow, the net force on the piston must be essentially zero, so the force $\vec{F}_{\text{gas}}$ that the gas exerts on the piston must be essentially equal in *magnitude* to the external force $\vec{F}_{\text{ext}}$ on the piston (although they have opposite directions): $\text{mag}(\vec{F}_{\text{ext}}) \approx \text{mag}(\vec{F}_{\text{gas}})$. Moreover, $\text{mag}(\vec{F}_{\text{gas}})$ in this case is equal to the pressure P that the gas exerts on the piston times the piston's area A. If we plug these results into equation T3.3, we get

$$dW \approx F_{\text{gas}} \, dr = PA \, dr \qquad (T3.4)$$

Note that the infinitesimal change in the gas's volume in this process is

$$dV \equiv V_{\text{final}} - V_{\text{initial}} = A(L_{\text{final}} - L_{\text{initial}}) = -A \, dr \quad \text{ for compression} \quad (T3.5)$$

This result is negative because the length of the cylinder has *decreased* by dr, meaning that the cylinder's volume has decreased. Plugging this into equation T3.4, we get

---

$$dW = -P \, dV \qquad (T3.6)$$

Purpose: This equation expresses the thermodynamic work dW done on a gas during a volume change dV.
Symbols: P is the gas pressure during the volume change.
Limitations: The volume change must be small enough that P does not change significantly during the process; and it must be slow enough that the forces on the piston are essentially in balance, the piston's kinetic energy is negligible, and the gas is in equilibrium with itself.
Note: To find the total work W done on the gas during a process in which the pressure varies, one must integrate; see section T3.5.

Work done by an infinitesimal slow volume change

The minus sign ensures that dW is positive (energy is flowing *into* the gas) for a compression even though dV is negative.

Exercise T3X.3

Consider the case of an infinitesimal *expansion*. Go through the steps outlined above again, and show that even though the signs of various equations along the way are different, you still end up with equation T3.6.

(a)   **(b)**   **(c)**

Figure T3.3
A ball thrown against the back of a truck will bounce back with (a) about the same energy as it had originally if the truck is at rest, (b) more energy if the truck is backing up, (c) less energy if the truck is moving forward.

A microscopic model for how energy is transformed

How exactly is energy being transferred to the gas in the case of a compression (or extracted from the gas in the case of an expansion)? We can understand the microscopic processes involved in such an expansion by using a simple analogy. Imagine bouncing tennis balls off the back end of a truck (see figure T3.3). If the truck is stationary and the balls are perfectly elastic, they will bounce back with about the same kinetic energy as they had before striking the truck. But if the truck is backing up toward you, the balls will bounce back from the truck with more energy relative to you than you gave them: some of the energy of the truck's motion is being converted to kinetic energy in the balls. Similarly, if the truck is going away from you, the balls will bounce back with less energy than they had to begin with: energy is being transferred from the balls to the truck.

In a similar way, gas molecules that bounce back elastically from a piston at rest will bounce back from an advancing piston (i.e., during compression) with more energy than they had originally, implying that the thermal energy of the gas increases at the expense of the piston's energy. When the piston retreats (during expansion), the molecules bounce back with less energy than they had originally, transferring the thermal energy of the gas to the motion of the piston.

T3.3   The State of a Gas

What properties of a sample of gas can we measure at the macroscopic level, that is, without measuring properties of individual molecules of this particular sample of gas? We can certainly measure the gas sample's pressure P, its volume V, and its temperature T without even knowing that molecules exist. We can also measure its total internal thermal energy U (at least in principle) by measuring how much energy we have to add to the gas to increase its temperature from absolute zero to its current temperature (while keeping its volume fixed). Finally, we can measure the gas sample's mass M, and if we know the gas's chemical composition, we can use this to calculate the number of moles and (given a value for Avogadro's number N_A) the number of molecules N in the sample. The quantities P, V, T, U, M, and N are, therefore, some of the **macroscopic properties** of the gas.

Note that these six quantities are *not* all independent. In chapter T2, we saw that these quantities are linked by the three equations

$$N = \frac{M}{M_A} N_A \qquad PV = Nk_B T \qquad \text{and} \qquad U = \frac{f}{2} Nk_B T \qquad \text{(T3.7)}$$

(Note that if we know the gas's chemical composition, in principle we know its molar mass M_A and the number of degrees of freedom f.) Therefore, if we know the sample's chemical composition and a suitably chosen *triplet* of properties, such as P, V, N or U, V, N, we can calculate the sample's other three macroscopic properties.

Exercise T3X.4

In each of the following cases, assume that you know the chemical composition of a sample of gas. (a) Given a triplet of values P, V, and N, how would you calculate U, T, and M? (b) Given the triplet U, V, and N, how you would calculate P, T, and M? (c) What triplet of values from the list P, V, T, U, M, and N would *not* provide you with enough information to calculate the other three values?

This means that two gas samples having the same chemical composition and the same values for a suitably chosen triplet of values will be physically indistinguishable if we are not allowed to examine the behavior of individual molecules. We say, therefore, that such a triplet of values describes the macroscopic state or **macrostate** of an ideal gas.

T3.4   *P–V* Diagrams and Constrained Processes

In many situations of interest, the number of molecules N in a sample of gas is known and fixed. When this is true, we need to know only two additional gas properties to completely determine its macrostate at a given instant of time. If we choose these variables to be P and V, then every macrostate of a fixed amount of gas corresponds to a unique *point* on a graph of P versus V, as shown in figure T3.4. We call such a graph a ***P–V* diagram.**

Since $PV = Nk_BT$, all the points on a *P–V* diagram corresponding to a given temperature lie on a curve such that $PV = $ constant, or $P \propto 1/V$. A set of such curves corresponding to various values of T is also shown in figure T3.4.

A **quasistatic gas process** is any process that changes the state of a gas and that is done slowly enough that the gas remains essentially in equilibrium at all times. This means that at all times the gas pressure P has the *same* well-defined value at all points in the gas. As the state of the gas changes, the point representing that state on the *P–V* diagram will move, marking out a *path* on the diagram. An example path is also shown in figure T3.4.

P and V are completely independent variables, and there is no intrinsic reason why a gas cannot have any pressure at any given volume. A process in principle (like the "general process" shown in the figure) might involve any connected sequence of points on the diagram. In practice, though, there are four special gas processes that keep coming up as useful approximations in realistic situations:

1. In **isochoric processes** the gas is heated or cooled while its volume is constrained to be constant (e.g., by keeping the gas in a rigid container). The root *iso-* means "same," and *choric* refers to volume.

Representing states and processes on a *P–V* diagram

Important (and useful) gas processes

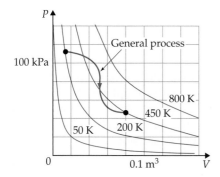

Figure T3.4

Every point on this *P–V* diagram represents a possible macroscopic state for a fixed amount of ideal gas. The thin curves connect gas states having the same temperature (the curves are labeled assuming that $N = 5.8 \times 10^{23}$). Any quasistatic process is an ordered succession of states and can be represented by a succession of points (i.e., a curve) on this diagram.

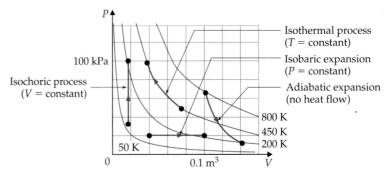

Figure T3.5

Curves on a *P–V* diagram corresponding to the four most useful types of processes. Again, for the values on the curves of constant temperature it is assumed that $N = 5.8 \times 10^{23}$.

2. In **isobaric processes** the gas is heated or cooled while its *pressure* is constrained to be constant (e.g., by confining the gas with a piston whose other side is acted on by the constant pressure of the earth's atmosphere). The root *bar-* refers to pressure (as in the word *barometer*).
3. In **isothermal processes** the gas is expanded or compressed while its temperature is constrained to be constant (e.g., by placing it in good thermal contact with a "bath" having a certain fixed temperature).
4. In **adiabatic processes** the gas is expanded or compressed while heat flow to or from the gas is constrained to be zero (e.g., by putting thermal insulation around the gas container). We will discuss adiabatic processes more fully in section T3.6.

These processes are illustrated in figure T3.5. We call these processes **constrained processes**, because the particular path on the *P–V* diagram that each process follows is determined by a *constraint* placed on the gas during the process (e.g., the gas in an isothermal process is constrained to have a constant temperature).

T3.5   Computing the Work

Calculating *W* when the pressure changes significantly during a process

These constrained processes are useful partly because we can actually compute the work done on or by the gas involved in such a process. As we noted before, the equation $dW = -P\,dV$ applies only to infinitesimal compressions or expansions. This is so because as a gas is compressed or expanded, its pressure *P* will generally change during the process. If we want to compute the total work *W* done during a volume change during which *P* changes significantly, we have to evaluate the integral:

$$W = -\int P\,dV \qquad\qquad (T3.8)$$

In words, this equation tells us that to compute the total work, we should divide the process into infinitesimal steps, compute $dW = -P\,dV$ for each step, and sum the result over all the steps.

Implications for particular constrained processes

To actually evaluate such an integral, we have to be able to express *P* during the process as a function of *V*. This is easy to do for each of the four special constrained processes listed in section T3.4 except for the adiabatic process, which we will consider in the next section. The *isochoric* process (where *V* is constant) is trivial: since the volume does not change ($dV = 0$), no work is done as the result of compression or expansion:

$$W = 0 \qquad\qquad (T3.9)$$

In an *isobaric* process, P is constant, so we can pull it out of the integral:

$$W = -\int P\,dV = -P\int_{V_i}^{V_f} dV = -P(V_f - V_i) = -P\,\Delta V \qquad (T3.10)$$

where V_i is the gas's initial volume and V_f is its final volume.

In an isothermal process, T is constant. The ideal gas law then implies that

$$PV = Nk_B T = \text{constant} = P_i V_i \quad \Rightarrow \quad P(V) = \frac{Nk_B T}{V} = \frac{P_i V_i}{V} \qquad (T3.11)$$

where P_i and V_i are the initial pressure and volume of the gas, respectively, and $P(V)$ means "pressure as a function of volume." Plugging this into equation T3.8, using $\int x^{-1}\,dx = \ln x$ and $\ln x - \ln y = \ln(x/y)$, you should be able to show that

$$W = -Nk_B T \ln \frac{V_f}{V_i} = -P_i V_i \ln \frac{V_f}{V_i} \qquad (T3.12)$$

Exercise T3X.5

Verify that equation T3.12 is correct.

Exercise T3X.6

A gas initially at atmospheric pressure (100 kPa) in a box 10 cm on a side is isothermally compressed to one-half that volume. What is W for this process?

One of the reasons that P–V diagrams of gas processes are so useful is that *the work done in a given quasistatic expansion or compression process is equal in magnitude to the area under the curve representing that process on a P–V diagram.* This is a direct consequence of equation T3.8: if we consider P to be a function of V, then the standard interpretation of the integral of $P(V)$ is that it corresponds to the area of the curve of P when it is plotted as a function of V (see figure T3.6).

Because of this simple visual interpretation of equation T3.8, one can get a lot of qualitative information about a process from a P–V diagram, as demonstrated in the following example.

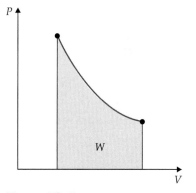

Figure T3.6
The *magnitude* of the work flowing into or out of a gas is equal to the area under the curve of P when plotted as a function of V. The sign of the work is positive if the gas is being compressed, negative if it expands.

Example T3.1

Problem   Imagine that an ideal monatomic gas is taken from its initial state A to state B by an *isothermal* process, from B to C by an *isobaric* process, and from C back to its initial state A by an *isochoric* process, as shown in figure T3.7a. Fill in the signs of Q, W, and ΔU (or zero if appropriate) for each step (assuming that no work is done through expansion or compression).

Solution   The finished chart is shown in figure T3.7b. These signs are determined as follows.

Process $A \to B$ is an isothermal expansion, meaning that T is constant. But for an ideal gas, U is proportional to T $[U = (f/2)Nk_B T]$, so ΔU is *zero* during an isothermal process. During an expansion, work energy flows *out* of the gas (remember that for each infinitesimal step along this process, $dW = -P\,dV$, and $dV > 0$ in an expansion process), so W is *negative*. But if

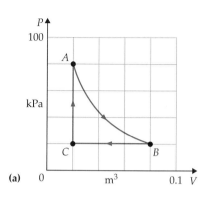

Figure T3.7
A sequence of gas processes.

Step:	Q	W	ΔU
$A \longrightarrow B$	$+$	$-$	0
$B \longrightarrow C$	$-$	$+$	$-$
$C \longrightarrow A$	$+$	0	$+$

(b)

work energy flows out of the gas and yet the total thermal energy U of the gas remains the same, then (in the absence of other kinds of work) heat energy must flow into the gas, so Q must be *positive*.

Process $B \rightarrow C$ is an isobaric *compression*, so W is *positive* here. On the other hand, the ideal gas law says that $PV = Nk_BT$. Therefore, since V is decreasing in this process while P remains constant, T must be decreasing (that is, $dT < 0$ for each small step in the process) which in turn implies that the thermal energy U of the gas must be decreasing: thus ΔU is *negative*. Since work energy is flowing *into* the gas and yet its thermal energy is *decreasing*, heat energy must be flowing *out* of the gas: Q is negative.

Finally, process $C \rightarrow A$ is an isochoric process. Since there is no change in volume, no work is done on or by the gas: $W = 0$. Yet the temperature is increasing (since PV is increasing), so the gas's thermal energy is increasing, meaning that ΔU is *positive* in this process. This increase in thermal energy must be supplied by heat (since $W = 0$), so Q is positive.

Estimating W from a P–V diagram

We can also estimate, directly from the diagram, the work energy flowing into or out of the gas in the cyclic process we've been considering. Each grid square's worth of area on the P–V diagram shown in figure T3.8 represents

$$(20 \times 10^3 \, \text{Pa})(0.020 \, \text{m}^3)\left(\frac{1 \, \text{N/m}^2}{1 \, \text{Pa}}\right)\left(\frac{1 \, \text{J}}{1 \, \text{N} \, \text{m}}\right) = 400 \, \text{J} \qquad \text{(T3.13)}$$

There are a total of about 6 squares (four whole squares, two squares mostly complete, and two small parts of squares) of area under the curve for the isothermal expansion $A \rightarrow B$, so the gas loses $6 \times 400 = 2400$ J of work energy during that process. In the isobaric compression $B \rightarrow C$, the gas gains $3 \times 400 = 1200$ J of work energy. Since $W = 0$ for the isochoric process, the total energy lost by the gas during the entire cyclic process is about 1200 J. Note that *the net work done by the gas in a cyclic process is equal to the area enclosed by the process*.

Note that since the gas comes back to the same state A (and thus same temperature) at the end of the process that it had originally, its thermal

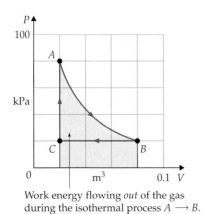

Work energy flowing *out* of the gas during the isothermal process $A \longrightarrow B$.

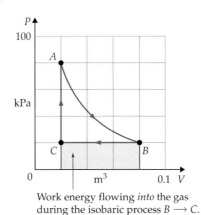

Work energy flowing *into* the gas during the isobaric process $B \longrightarrow C$.

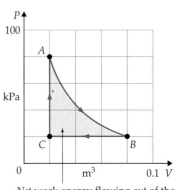

Net work energy flowing *out* of the gas during the entire cyclic process.

Figure T3.8
Net work energy flow during a cyclic process.

energy must be the same at the end of the cycle as it was in the beginning (since U for an ideal gas only depends on N and T). Yet the gas loses about 1200 J of work energy in the cycle, as we have just seen. Where does this energy come from, if not from the thermal energy of the gas? It must come from the *heat* energy that we put into the gas. This cycle is therefore an example of a process that converts heat energy to work energy. Many kinds of heat engines use such cyclic processes to produce useful work energy from heat energy. We will talk more about heat engines in chapter T9.

Exercise T3X.7

The figure below shows another cyclic gas process. Fill out the chart as we did in example T3.1. Also estimate the net amount of work energy flowing out of the gas in this process. In which step of the process was energy supplied in the form of heat?

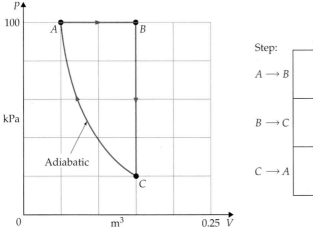

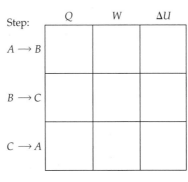

Step:	Q	W	ΔU
$A \longrightarrow B$			
$B \longrightarrow C$			
$C \longrightarrow A$			

Exercise T3X.8

Check my estimate for the work energy lost during process $A \rightarrow B$ in figure T3.8, using the exact formula T3.12. Is our estimate within about 10% or so of the exact answer? (10% is about the best one can do by eye.)

T3.6   Adiabatic Processes

Exercise T3X.8 included an *adiabatic* compression process as one of the steps in the cyclic process. In an adiabatic process, no heat is permitted to enter or leave the system in question. (*Adiabatic* comes from Greek *adiabatos*, meaning "impassable": the connotation is that the system's boundary is impassable to heat.) We have already seen that during an isothermal process, T is constant (by definition) and $P \propto 1/V$. How do T and P depend on V in an adiabatic process? Our goal in this section is to use the microscopic model of an ideal gas to find out.

Consider a gas of N molecules confined in a cylinder by a piston whose surface area is A (see figure T3.9a). Imagine that at a given instant, the length of the cylinder is L and the piston is moving inward (the $-x$ direction) at a speed u which is much smaller than the average thermal speed of the molecules.

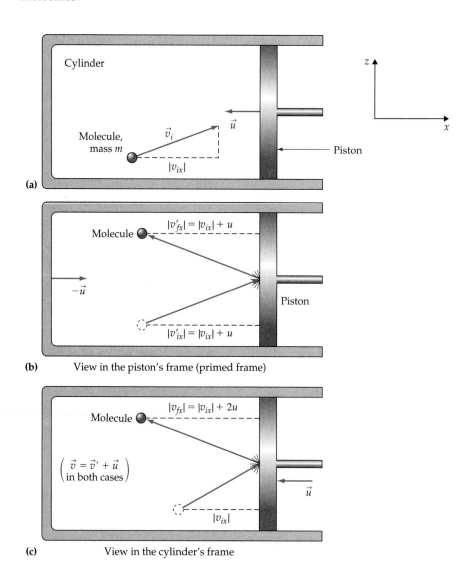

Figure T3.9

(a) A molecule of mass m in a cylinder of gas approaches a piston that is moving inward. (b) In the piston's frame, the molecule rebounds with the same speed that it had initially. (c) In the cylinder's frame, the molecule rebounds with a *higher* speed than it had initially.

In section T3.2, I described qualitatively why a molecule bouncing off a moving piston gains energy from that piston. Let us see if we can determine exactly how much energy the molecule gains. Imagine a molecule of mass m approaching the piston with an initial x-velocity whose magnitude is $|v_{ix}|$ in the frame of the cylinder. Since the piston is moving toward the molecule with speed u, the molecule's initial x-velocity *relative to the piston* will be $|v_{ix}| + u$. If we assume the collision is elastic, the molecule should rebound from the piston with the same speed *relative to the piston* as it had initially, meaning that its x-velocity relative to the piston will still have the magnitude $|v_{ix}| + u$ (see figure T3.9b). But since the piston is moving with speed u in the same direction, the molecule's final x-velocity relative to the *cylinder* has magnitude $|v_{fx}| = (|v_{ix}| + u) + u = |v_{ix}| + 2u$ (see figure T3.9c). Since the y and z components of the molecule's velocity are not affected by the collision, the change in the molecule's kinetic energy during this collision is therefore

How T depends on V during an adiabatic compression

$$\Delta K = K_f - K_i = \tfrac{1}{2}m(v_{fx}^2 + v_{fy}^2 + v_{fz}^2) - \tfrac{1}{2}m(v_{ix}^2 + v_{iy}^2 + v_{iz}^2) = \tfrac{1}{2}m(v_{fx}^2 - v_{ix}^2)$$

$$= \tfrac{1}{2}m(|v_{ix}| + 2u)^2 - \tfrac{1}{2}m|v_{ix}|^2 = 2mu|v_{ix}| + 2mu^2 \approx 2mu|v_{ix}| \qquad \text{(T3.14)}$$

The final approximation is valid because we are assuming that $u \ll |v_{ix}|$. This assumption also means that the piston does not move much between collisions, so the time that has passed since the previous time the molecule hit the piston is still very nearly $\Delta t \approx 2L/|v_{ix}|$. Thus the rate at which this particular molecule's kinetic energy increases is

$$\frac{dK}{dt} \approx \frac{\text{kinetic energy gain per collision}}{\text{time between collisions}} = \frac{2mu|v_{ix}|}{2L/|v_{ix}|} = \frac{muv_{ix}^2}{L} \qquad \text{(T3.15)}$$

To find the rate at which the gas's total thermal energy increases, we sum this over all molecules:

$$\frac{dU}{dt} \approx \sum_{j=1}^{N} \frac{um_j v_{j,ix}^2}{L} = \frac{Nu}{L} \frac{1}{N} \sum_{j=1}^{N} m_j v_{j,ix}^2 \equiv \frac{Nu}{L}[m_j v_{j,ix}^2]_{\text{avg}} = \frac{Nu}{L} k_B T \qquad \text{(T3.16)}$$

where the last step follows from equation T2.6.

Now, for a ideal *monatomic* gas,

$$U = \frac{3}{2} N k_B T \qquad \Rightarrow \qquad \frac{dU}{dt} = \frac{3}{2} N k_B \frac{dT}{dt} \qquad \text{(T3.17)}$$

Plugging the last result into equation T3.16, we get

$$\frac{3}{2} N k_B \frac{dT}{dt} \approx \frac{N k_B T}{L} u \qquad \Rightarrow \qquad \frac{1}{T} \frac{dT}{dt} \approx \frac{2}{3} \frac{u}{L} \qquad \text{(T3.18)}$$

Now, $u = -dL/dt$ (when u is positive, L is decreasing, so dL/dt is negative). Therefore

$$\frac{u}{L} = -\frac{1}{L} \frac{dL}{dt} = -\frac{1}{AL} \frac{d(AL)}{dt} = -\frac{1}{V} \frac{dV}{dt} \qquad \text{(T3.19)}$$

Plugging this into the previous equation, we get

$$\frac{1}{T} \frac{dT}{dt} = -\frac{2}{3}\left(\frac{1}{V} \frac{dV}{dt}\right) \qquad \Rightarrow \qquad 0 = \frac{1}{T} \frac{dT}{dt} + \frac{2}{3}\left(\frac{1}{V} \frac{dV}{dt}\right) \qquad \text{(T3.20)}$$

If we multiply both sides of the rightmost expression by $TV^{2/3}$ (we will see why shortly), we get

$$0 = V^{2/3} \frac{dT}{dt} + \frac{2}{3} \frac{T}{V^{1/3}} \frac{dV}{dt} = \frac{d}{dt}(TV^{2/3}) \qquad \text{(T3.21)}$$

Exercise T3X.9

Verify that if you take the time derivative of $TV^{2/3}$, using the chain and product rules, you get the quantity to left of the rightmost equality.

Why this is physically sensible

This last result means that $TV^{2/3}$ is *constant* for a monatomic gas undergoing an adiabatic volume change! Qualitatively, we already knew that T had to increase during an adiabatic compression (work is being done on the gas, but heat cannot flow out, so U and thus T must increase), but this relation spells this out quantitatively: $T \propto 1/V^{2/3}$ for a monatomic gas during an adiabatic volume change.

The relationship between P and V

Since $T = PV/Nk_B$ according to the ideal gas law, equation T3.21 implies that

$$0 = \frac{d}{dt}(TV^{2/3}) = \frac{d}{dt}\left(\frac{PV}{Nk_B}V^{2/3}\right)$$

$$= \frac{1}{Nk_B}\frac{d}{dt}(PV^{5/3}) \quad \Rightarrow \quad PV^{5/3} = \text{const} \quad (T3.22)$$

Note that in an isothermal volume change, $P \propto 1/V$. Here we see that $P \propto 1/V^{5/3}$ for a monatomic gas during an adiabatic volume change. This also makes good physical sense. During an *isothermal* compression the pressure increases as $1/V$ because reducing the volume proportionally reduces the molecular transit time between collisions, which proportionally *increases* the collision rate and thus the pressure. In an adiabatic compression, this is still true, but in addition the gas *temperature* increases during the compression. This increases the average molecular speed, which not only further decreases the transit time but also makes each collision more violent. So the pressure *should* increase more rapidly than $1/V$ during an adiabatic compression.

Generalizing to nonmonatomic gases

Equations T3.21 and T3.22 apply only to monatomic gases, but you can easily generalize the derivation to other kinds of gases as well.

Exercise T3X.10

For ideal gases in general, $U = (f/2)Nk_BT$ with $f = 3$ for a monatomic gas and $f \geq 5$ for other kinds of gases. Find the place in the previous derivation where I assumed that $f = 3$, and rework the derivation from that point to handle arbitrary values of f. You should find that equations T3.21 and T3.22 become $TV^{2/f} = \text{constant}$ and $PV^{(1+2/f)} = \text{constant}$, respectively.

Physicists usually write these laws in the following form:

$$TV^{\gamma-1} = \text{constant} \qquad (T3.23a)$$

$$PV^{\gamma} = \text{constant} \qquad (T3.23b)$$

Purpose: These equations specify how a gas's temperature T and pressure P vary with its volume V during an adiabatic process.
Symbols: $\gamma = 1 + 2/f$ is the gas's **adiabatic index,** where f is the number of molecular degrees of freedom for the gas.

Limitations: The process must be adiabatic, the gas must be ideal, and the compression must occur very slowly compared to the thermal speed of the gas molecules.

Note: The values of γ for various types of gases are as follows:

$$\text{Monatomic:} \quad \gamma = \frac{5}{3} = 1.67$$

$$\text{Diatomic:} \quad \gamma = \frac{7}{5} = 1.40$$

$$\text{Polyatomic:} \quad 1 < \gamma \leq \frac{4}{3} = 1.33$$

These equations have a host of practical applications, and they will be very useful to us in chapters T8 and T9.

Example T3.2

Problem   Imagine that we compress a sample of air whose initial pressure is 100 kPa and temperature is 22°C (= 295 K) to a volume that is one-quarter of its original volume within 0.1 second (s). The piston in this case moves a total distance of 30 cm. What is the final temperature of the gas?

Model   The average speed of the piston during this compression is about 3 m/s, which is much smaller than the typical speeds of air molecules at room temperature (recall that $v_{\text{rms}} = 520$ m/s for nitrogen at room temperature). On the other hand, even if the cylinder is not very well insulated, the process occurs so quickly that not much heat will be able to flow into or out of the cylinder. Therefore, we can model this as an adiabatic compression. Since air is a diatomic gas and is reasonably ideal at normal pressures and temperatures, it should obey the equation $TV^{\gamma-1} = \text{constant}$ with $\gamma = 1.4$.

Solution   In this case, therefore,

$$T_f V_f^{\gamma-1} = T_i V_i^{\gamma-1} \quad \Rightarrow \quad T_f = T_i \left(\frac{V_i}{V_f} \right)^{\gamma-1}$$

$$= 295\,\text{K} \left(\frac{V_i}{\frac{1}{4}V_i} \right)^{0.4} = 514\,\text{K} \quad \text{(T3.24)}$$

TWO-MINUTE PROBLEMS

T3T.1   In each of the processes described below, energy is being transformed to or from thermal energy. Does the energy flow involve heat (T) or work (F)?
   a. Your car's brakes get hot when used repeatedly.
   b. Your pizza gets warm in a microwave oven.
   c. An electric stove element gets hot when turned up.
   d. Your car gets hot in the sun on a relatively cool day.
   e. You get cooler when standing in the breeze from a fan.

T3T.2 A gas with a pressure of 100 kPa is in a container that is a cube measuring 10 cm on a side. If we move one wall in 1 millimeter (mm), does work flow into (A) or out of (B) the gas? What is the magnitude of W?
A. 1000 J
B. 100 J
C. 10 J
D. 1 J
E. 0.001 J
F. Other (specify)

Problems T3T.3 through T3T.7 refer to the following cyclic gas process:

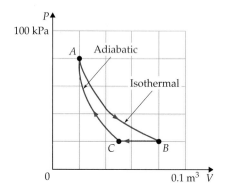

T3T.3 The process $B \to C$ shown is
A. An isochoric process
B. An isothermal process
C. An isobaric process
D. An adiabatic process
E. An isometric process
F. None of the above

T3T.4 What are the signs of Q, W, and ΔU for the process $A \to B$?
A. 0, −, −
B. 0, +, +
C. +, −, 0
D. +, +, 0
E. −, +, 0
F. Other (specify)

T3T.5 What are the signs of Q, W, and ΔU for the process $B \to C$? (Select from the answers provided in problem T3T.4.)

T3T.6 What are the signs of Q, W, and ΔU for the process $C \to A$? (Select from the answers provided in problem T3T.4.)

T3T.7 Is the work energy flowing into or out of the gas in process $B \to C$ positive (A) or negative (B)? Which of the values below is closest to the magnitude of W?
A. 0.6 J
B. 1.5 J
C. 300 J
D. 600 J
E. 1500 J
F. 3000 J

T3T.8 Imagine that a bubble of helium (a monatomic gas) rising from the bottom of the ocean expands in volume by a factor of 8 by the time it reaches the surface [where the pressure is 1 atmosphere (atm)]. Assume that the bubble rises so fast that it expands essentially adiabatically. What was the pressure on the gas at the depth where it formed (in atmospheres)?
A. 3.5 atm
B. 8 atm
C. 16 atm
D. 18 atm
E. 32 atm
F. Other (specify)

T3T.9 If the temperature of the bubble described in problem T3T.8 was 320 K when it formed, what is its approximate final temperature when it reaches the surface?
A. 40 K
B. 80 K
C. 320 K
D. 1280 K
E. 2560 K
F. Other (specify)

HOMEWORK PROBLEMS

Basic Skills

T3B.1 A gas is confined to a cylinder by a piston. The gas has an initial pressure of 120 kPa and a volume of 100 cm³. The piston is slowly moved back until the gas's volume has increased by 0.5%. What is the approximate work that flows into or out of the gas? (Be sure to give the correct sign as well as the correct magnitude. You can consider this volume change to be small enough that P is approximately constant during the process.)

T3B.2 A gas is confined to a cylinder by a piston. The gas has an initial pressure of 95 kPa and a volume of 300 cm³. The piston is slowly pushed in until the gas's volume has decreased by 1%. What is the approximate work that flows into or out of the gas? (Be sure to give the correct sign as well as the correct magnitude. You can consider this volume change to be small enough that P is approximately constant during the process.)

T3B.3 An ideal gas in a cylinder is allowed to expand while its temperature is held fixed. Is there any heat flow involved in this process? If so, does heat energy flow into or out of the gas? Please explain.

T3B.4 An ideal gas in a cylinder is slowly compressed while its pressure is held fixed. Is there any heat flow involved in this process? If so, does heat energy flow into or out of the gas? Please explain.

T3B.5 An ideal gas with an initial pressure of 120 kPa is confined to a cylinder with a volume of 150 cm³. We then allow it to slowly expand to a volume of 350 cm³ while adding enough heat to keep its pressure fixed. What is the work that flows into or out of the gas (be sure to give the correct sign as well as magnitude)?

T3B.6 A monatomic ideal gas with an initial pressure of 80 kPa is confined to a cylinder with a volume of 600 cm³. We then compress the gas isothermally until its volume has decreased to 450 cm³. What is its pressure now?

T3B.7 A monatomic ideal gas with an initial pressure of 60 kPa is confined to a cylinder with a volume of 600 cm³. We then compress the gas adiabatically until its volume has decreased to 450 cm³. What is its pressure now?

Synthetic

T3S.1 A friend observes while heating water over a flame that the temperature of the water does not increase after it begins boiling. Your friend comments, "I guess that heat is no longer flowing from the flame to the water." You state that you think heat is still flowing into the water. Your friend responds, "But heat is energy flow associated with a temperature difference, and the water's temperature is constant, so there is no temperature difference." Gently explain to your friend why his or her statement is incorrect.

T3S.2 One mole of helium gas in a cylinder is allowed to expand adiabatically. During the expansion its temperature falls from 310 K to 265 K. How much work energy flows in this expansion? (Be sure to give both the correct magnitude and the correct sign.)

T3S.3 Some 3.0×10^{22} molecules of nitrogen gas at 280 K are constrained to expand isothermally to 3 times the original volume. Heat must enter the gas during this process. Why? How much heat enters the gas in this process?

T3S.4 A gas is constrained to follow the three-step cyclic process shown in the graph below. Prepare a chart (like the one shown in example T3.1) that specifies the sign of Q, W, and ΔU for each step in the process. What is the net work flowing into or out of the gas for the entire cyclic process (be sure to give the correct magnitude and sign)?

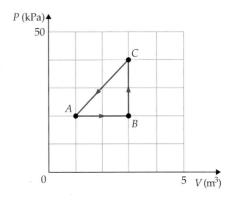

T3S.5 A gas is constrained to follow the three-step cyclic process shown in the graph below. Prepare a chart (like the one shown in example T3.1) that specifies the sign of Q, W, and ΔU for each step in the process.

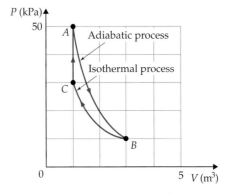

T3S.6 Heat must flow into the gas during the process $C \rightarrow A$ shown in the drawing associated with problem T3S.5. Why? If the gas's temperature is 290 K at point C, find its temperature at point A and the heat that has flowed into the gas in this process if the gas is monatomic.

T3S.7 A bubble of air is formed at the bottom of the ocean floor 66 ft below the surface, where the ambient pressure is about 300 kPa = 3 atm. The bubble has an initial volume of about 25 cm³ and a temperature of 8°C. If the bubble rises so fast that it expands essentially adiabatically, what is its final volume? What is its final temperature?

T3S.8 A research balloon bound for the stratosphere (see figure T3.10) is filled at sea level with 800 m³ of helium whose initial temperature is 285 K. The balloon is released, and it climbs to an altitude where the air pressure is 0.045 times its sea-level value. If the helium expands adiabatically, what is the balloon's volume now? What is the temperature of the helium?

Figure T3.10
(for problem T35.8.)

T3S.9 Use equation T3.23 to show that the work done during an adiabatic volume change from V_i to V_f is

$$W = \frac{P_i V_i^\gamma}{\gamma - 1}\left(\frac{1}{V_f^{\gamma-1}} - \frac{1}{V_i^{\gamma-1}}\right) \qquad (T3.25)$$

(*Hint:* Solve equation T3.23b for P as a function of V, and then use equation T3.8.)

T3S.10 This problem outlines a way to derive equations T3.23 without delving into the microscopic behavior of molecules. (This method is marginally simpler than the method in the text but is more abstract and yields less physical insight.)

(a) Consider an infinitesimal adiabatic volume change dV. During this process, the gas's temperature will change by some tiny amount dT, and the pressure will change by some tiny amount dP. Show that if you take the derivative of the ideal gas law with respect to T and multiply both sides by dT, you get

$$P\,dV + V\,dP = Nk_B\,dT \qquad (T3.26)$$

(b) By considering how dW relates to dU in this case, argue that

$$P\,dV = -\frac{f}{2}Nk_B\,dT \qquad (T3.27)$$

(c) Use this to eliminate $Nk_B\,dT$ from equation T3.26 and rearrange things to show that

$$0 = \frac{dP}{dV} + \gamma\frac{P}{V} \quad \text{where} \quad \gamma \equiv 1 + \frac{2}{f} \qquad (T3.28)$$

(d) Multiply both sides by an appropriate power of V and show that

$$0 = \frac{d}{dV}(PV^\gamma) \quad \Rightarrow \quad PV^\gamma = \text{constant} \qquad (T3.29)$$

(e) Use the ideal gas law to show from this that $TV^{\gamma-1} = \text{constant}$.

Rich-Context

T3R.1 Imagine that the atmospheric pressure at the top of a tall mountain is about 0.65 times the pressure at sea level. If it is 30°C (86°F) at the beach and a stiff breeze blows this air up the mountain so rapidly that the air essentially expands adiabatically, what is the approximate temperature at the top of the mountain?

What is the temperature at the top?

T3R.2 Imagine that we have a sample of nitrogen gas confined to a cylinder by a movable piston. The cylinder is immersed in ice water, and the gas initially has a temperature of 0°C. (1) You fairly rapidly compress the gas, adiabatically decreasing its volume by a factor of 2. (2) You hold the piston still until the gas has cooled again to 0°C, and then (3) you allow the gas to expand slowly to its original volume while allowing plenty of time for heat to move out of or into the gas (so that its temperature remains very close to 0°C). If 85 g of ice was melted during the second step of the cycle, how much work energy did you put into the gas during the first step? Describe your reasoning. (*Hint:* I suggest that you first draw a *P–V* diagram. Also, it only *looks* as if you do not have enough information to solve this problem. If you find yourself doing lots of calculations, stop and *think* about it some more.)

Advanced

T3A.1 One can find the work involved in an adiabatic process in one of two ways. The first way is to use equation T3.25. The second way is to realize that since no heat flows into the gas, the work that flows into (or out of) the gas is the same as the gas's change in internal energy: $\Delta U = W$. You can find ΔU by using equation T3.23a and the results of the last chapter. Prove mathematically that this method yields the same results as equation T3.25.

ANSWERS TO EXERCISES

T3X.1 (a) Q, (b) Q, (c) W, (d) Q, (e) W, (f) Q, (g) Q, (h) W.

T3X.2 (a) K, (b) W, (c) W, (d) K, (e) W, (f) W.

T3X.3 In this case, you still need to apply an inward external force to the piston to keep the gas confined, but now you let the piston move outward. This means that $\vec{F}_{ext}$ and $d\vec{r}$ are opposite, so the work done in each infinitesimal step is $dW = \vec{F}_{ext} \cdot d\vec{r} = -F_{ext}\, dr$. We still have $F_{ext} \approx F_{gas} = PA$, but now $dV = A(L_{final} - L_{initial}) = +A\, dr$. So $dW = -PA\, dr = -P\, dV$, as before.

T3X.4 (a) Given P, V, and N, we can calculate M, T, and U as follows:

$$M = M_A \frac{N}{N_A} \qquad T = \frac{PV}{Nk_B}$$

and $\qquad U = \frac{f}{2} N k_B T \qquad$ (T3.30)

(Knowing the chemical composition of the gas gives you M_A and n.)

(b) Given U, V, and N, we can calculate M, T, and P as follows:

$$M = M_A \frac{N}{N_A} \qquad T = \frac{2}{f} \frac{U}{N k_B}$$

and $\qquad P = \frac{N k_B T}{V} \qquad$ (T3.31)

(c) A triplet that *does not* work is U, N, and T; we cannot determine P or V from this information. Also, any triplet containing both M and N will not work, because knowing M_A enables one to directly calculate M from N or vice versa (so specifying both gives one no more information than specifying either one). All other triplets work.

T3X.5 Substituting as suggested, we get

$$W = -\int_{V_i}^{V_f} \frac{N k_B T}{V}\, dV = -N k_B T \int_{V_i}^{V_f} \frac{dV}{V}$$

$$= -N k_B T (\ln V_f - \ln V_i) = -N k_B T \ln \frac{V_f}{V_i} \quad (T3.32)$$

According to the ideal gas law, $P_i V_i = N k_B T = P_f V_f$, so we can substitute either $P_i V_i$ or $P_f V_f$ for $N k_B T$ in equation T3.32.

T3X.6 Plugging $V_f = \frac{1}{2} V_i$ into equation T3.12 yields

$$W = -P_i V_i \ln \frac{V_f}{V_i}$$

$$= -\left(100{,}000 \frac{N}{m^2}\right)(0.10\, m)^3 \left(\frac{1\, J}{1\, N\, m}\right) \ln \frac{1}{2}$$

$$= +69\, J \qquad (T3.33)$$

T3X.7 The chart should look like this:

Step:	Q	W	ΔU
$A \rightarrow B$	+	−	+
$B \rightarrow C$	−	0	−
$C \rightarrow A$	0	+	+

There is more work flowing out of the gas during step $A \rightarrow B$ than into the gas in step $C \rightarrow A$, so the net work is negative. The area inside the cycle in the diagram is about 5.5 squares. Each square corresponds to an energy of

$$\left(20{,}000 \frac{N}{m^2}\right)(0.05\, m^3)\left(\frac{1\, J}{1\, N\, m}\right) = 1000\, J \qquad (T3.34)$$

so $W \approx -5500$ J.

T3X.8 According to equation T3.12, the work involved in the expansion $A \rightarrow B$ is

$$W = -P_i V_i \ln \frac{V_f}{V_i}$$

$$= -\left(80{,}000 \frac{N}{m^2}\right)(0.02\, m^3)\left(\frac{1\, J}{1\, N\, m}\right) \ln \frac{0.08\, m^3}{0.02\, m^3}$$

$$= -2220\, J \qquad (T3.35)$$

I had estimated that the gas would lose 2400 J during this process, so my estimate was too high by about 8%.

T3X.9 According to the product and chain rules,

$$\frac{d}{dt}(TV^{2/3}) = V^{2/3} \frac{dT}{dt} + T \frac{d}{dt}(V^{2/3})$$

$$= V^{2/3} \frac{dT}{dt} + \frac{2}{3} T V^{-1/3} \frac{dV}{dt} \qquad (T3.36)$$

This is indeed the same as the rightmost quantity in equation T3.20.

T3X.10 Everything in the derivation applies to all ideal gases up to equation T3.17. We should substitute $f/2$ for $3/2$ in this equation and equation T3.18. This means that we should change the factor of $2/3$ appearing in equations T3.18, T3.20, and T3.21 to $2/f$, so equation T3.21 reads

$$0 = \frac{d}{dt}(TV^{2/f}) \quad \Rightarrow \quad TV^{2/f} = \text{constant} \quad (T3.37)$$

Since $T = PV/Nk_B$ according to the ideal gas law,

$$(PV)V^{2/f} = PV^{1+2/f} = (\text{constant})\,(Nk_B)$$
$$= \text{a different constant} \qquad (T3.38)$$

as claimed.

T4

Macrostates and Microstates

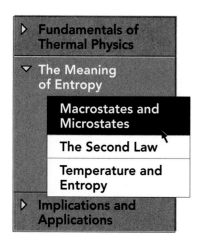

Chapter Overview

Introduction

The ideal gas model we explored in chapters T2 and T3 is useful in many ways, but it does not really help us understand why some processes are irreversible. We will discuss this problem in the next three chapters. This particular chapter introduces the crucial concept of *multiplicity* and develops a model for a monatomic solid that we will use throughout the discussion.

Section T4.1: Distinguishing Macrostates from Microstates

We describe a system's **macrostate** by listing its macroscopic properties, but its **microstate** by specifying the quantum state of every molecule in the system. In principle we can calculate a system's macroscopic properties (and thus its macrostate), knowing its microstate. In general, there are an immense number of microstates corresponding to any given macrostate; for example, we can distribute energy in many ways among an object's atoms without affecting its macroscopic total energy U.

Section T4.2: The Einstein Model of a Solid

We can model a monatomic solid by imagining that each of its atoms independently oscillates around its equilibrium position. Mathematically, we can treat an atom's oscillation in three dimensions as three independent one-dimensional oscillations, one along each coordinate axis. Since the atoms are identical and the coordinate axes equivalent, we can model an N-atom solid as being $3N$ identical but independent one-dimensional quantum oscillators. We call a solid that is adequately described by this simple model an **Einstein solid.**

According to unit Q, a one-dimensional quantum oscillator has evenly spaced energy levels separated by $\varepsilon \equiv \hbar(k_s/m)^{1/2}$, where k_s is the effective spring constant of the interactions holding an atom to its equilibrium position, m is the atom's mass, and $\hbar$ is Planck's constant over 2π. If we define our zero of energy so that the oscillator's ground state corresponds to zero energy, then the energy of any one of our $3N$ oscillators is simply an integer multiple of ε. An Einstein solid's total thermal energy is therefore

$$U = \sum_{i=1}^{3N} \varepsilon n_i \qquad \text{(T4.5)}$$

Purpose: This equation specifies an Einstein solid's total thermal energy U.

Symbols: $n_1, n_2, n_3, \ldots$ are a set of nonnegative integers (three per atom), and ε is the fixed difference between energy levels of each oscillator.

Limitations: The Einstein model only works well for monatomic solids at temperatures above 100 K or so (the exact limit depends on the solid).

Section T4.3: Counting Microstates

If we assume that an Einstein solid's volume is fixed, its macrostate is adequately described by its total thermal energy U and the number of atoms N. The **multiplicity** Ω of such a macrostate is therefore simply the number of different ways we can add $3N$ integer multiples of ε to get U. The general formula (see problem T4S.8) turns out to be

$$\Omega(N, U) = \frac{(q + 3N - 1)!}{q!(3N - 1)!} \tag{T4.6}$$

Purpose: This equation specifies the multiplicity Ω of an Einstein solid's macrostate.

Symbols: N is the number of atoms in the solid; U is its total energy; $q \equiv U/\varepsilon$ is the number of energy units to be distributed among the atoms; $n!$ or **n factorial** $\equiv 1 \cdot 2 \cdot 3 \cdots (n-1) \cdot n$.

Limitations: This equation only applies to an Einstein solid.

Section T4.4: Two Einstein Solids in Thermal Contact

Imagine now that we bring two Einstein solids A and B into thermal contact but isolate them from everything else (so that the combined system's total energy is fixed). We describe a **macropartition** of the combined system by specifying the macrostate of each solid. A **macropartition table** lists all the combined system's macropartitions and their multiplicities. Table T4.1 is a macropartition table for the case where $N_A = N_B = 1$ and the combined system's total energy is $U = U_A + U_B = 6\varepsilon$.

Note that the combined system's multiplicity Ω_{AB} in a given macropartition is the *product* of the multiplicities of each subsystem, because for each one of the Ω_A possible microstates for solid A, there are Ω_B possible microstates for solid B consistent with the macropartition.

Section T4.5: The Fundamental Assumption

The **fundamental assumption** of statistical mechanics is that *all a system's accessible microstates are equally likely in the long run*. This means that as the solids randomly exchange energy, a macropartition embracing many microstates is more probable than one embracing only a few. Thus we are likely to see the $3\!:\!3$ macropartition in table T4.1 about $100/28 = 3.6$ times more often than either the $0\!:\!6$ or $6\!:\!0$ macropartitions.

Section T4.6: Using StatMech

Creating tables by hand for systems having more than a handful of atoms and/or energy units becomes very tedious, but a computer can do the necessary calculations rapidly and accurately. The program StatMech (which you can download from the *Six Ideas* website) constructs such tables for larger systems.

Table T4.1   Possible macropartitions for $N_A = N_B = 1$, $U = 6\varepsilon$

Macropartition	U_A	U_B	Ω_A	Ω_B	Ω_{AB}
0:6	0	6	1	28	28
1:5	1	5	3	21	63
2:4	2	4	6	15	90
3:3	3	3	10	10	100
4:2	4	2	15	6	90
5:1	5	1	21	3	63
6:0	6	0	28	1	28

Grand total number of microstates $= 462$

T4.1   Distinguishing Macrostates from Microstates

In this chapter, we begin exploring Boltzmann's solution to the puzzle of irreversibility. The first step in understanding this solution is to understand the crucial distinction between the *macrostate* and the *microstate* of a thermodynamic system.

Definition of *macrostate*

The **macrostate** of a thermodynamic system is that system's state as characterized by its *macroscopically measurable properties*. We say, for example, that a system has the same macrostate at two different instants of time if and only if *all* its macroscopically measurable properties are the same at those two instants. We can completely describe a system's macrostate by specifying some minimal set of macroscopic properties that suffice to calculate all its other properties. For example, we saw in chapter T3 that we can completely describe an ideal gas's macrostate by specifying its chemical composition and values for its thermal energy U, its volume V, and the number of molecules N.

Definition of *microstate*

A system's **microstate,** on the other hand, is characterized by describing the state of *each individual molecule* in the system at a given time. In the context of newtonian mechanics, we might characterize a structureless particle's state by describing its position $[x, y, z]$ and velocity $[v_x, v_y, v_z]$: if we know these six quantities (as well as the molecular mass m), we could calculate any other quantity of interest for the particle (such as its kinetic energy, its potential energy relative to any other particle, and so on). So to specify the newtonian microstate of a monatomic system containing N identical atoms, we have to specify a minimum of $6N$ numbers. Note that while this idea is *conceptually* straightforward, there are so many atoms in even the tiniest speck of substance that it would be impossible in *practice* to measure these numbers at a given time. (Even simply writing them down would require more time than the projected lifetime of the sun.)

Actually, it is better to describe the state of something as small as a molecule using *quantum* mechanics instead of newtonian mechanics. In quantum mechanics, we describe the state of a given quanton in a system by specifying the quanton's energy level in the context of a certain model of the system. For example, we might model a molecule in a monatomic ideal gas as being a "quanton in a box" (in the language of unit Q), where the "box" is the container holding the gas. As we saw in unit Q, we can specify the energy state of a quanton moving in a *one*-dimensional box by a single integer n specifying how many half-wavelengths of the quanton's wave function fit between the box boundaries. To describe a quanton's quantum state in a *three*-dimensional box, we need to specify *three* integers n_x, n_y, and n_z (which describe how many half-wavelengths of the quanton's wave function fit between the box walls along the x, y, and z axes, respectively). So to describe the microstate of an ideal monatomic gas according to the box model, we need to list $3N$ numbers, three for every atom.

Note that if we know the microstate of a given thermodynamic system, we know its macroscopic properties as well. For example, imagine we have a sample of a monatomic ideal gas whose molecules have mass m, and we are given a list of the newtonian positions and velocities of all its molecules. By counting the numbers in the list, we can determine N, and the volume of the smallest box enclosing all the molecular positions gives us V. We can also calculate the kinetic energy of each molecule, and then add the results to find the total thermal energy U of the gas. This is enough information to specify the sample's macrostate.

There are *many* microstates in a macrostate

The most important thing to understand about microstates and macrostates is that a system in a given, well-specified macrostate could be in any one of a *huge* number of different microstates that we are unable to distinguish by macroscopic measurements. For example, imagine we describe the macrostate of an ideal monatomic gas by stating values for U, V, and N. There are a myriad of possible microstates that nonetheless add up to the same total U (each simply corresponds to a different way of distributing that total energy among the molecules). For any complex system, there are an immense number of possible microstates consistent with any given macrostate.

Perhaps the following analogy will make these ideas more vivid. Consider your room. It has two fundamental "macrostates" that a person (say, your parent) can rapidly discern without much detailed examination: "clean" or "messy." Describing your room's microstate, on the other hand, would involve meticulous documentation of the exact position and orientation of every object in the room. Now, there are probably a fairly large number of arrangements of objects in your room that your parent might consider clean (e.g., there are a number of possible ways to arrange your socks in the dresser drawer). There are vastly more possibilities for object arrangements that your parent would consider messy (just imagine the number of ways you could distribute your socks on the floor!). Either way, though, there are many microstates in a macrostate.

T4.2   The Einstein Model of a Solid

To actually *count* how many microstates there are in a macrostate, we need to talk about a specific model of a system. For the remainder of this chapter (and chapter T5), I will focus on a particular simplified model for a monatomic solid.

In 1907, Albert Einstein published a paper that proposed a simple but reasonably accurate model for predicting the thermal behavior of monatomic solids (such as crystals of pure aluminum, copper, carbon, or gold). Einstein proposed treating the atoms in such a solid as if they were held in their lattice position by springs, as illustrated in figure T4.1. In a real solid, the atoms are actually held in place by interatomic electromagnetic interactions. However, as we saw in unit C, the potential energy functions for such interactions become approximately the same as those for a mass on a spring in the small-oscillation limit. So in this model, we will assume that each atom's potential energy when it is a distance r from its equilibrium position is $\frac{1}{2}k_s r^2$ independent of the direction of r, just as it would be if it were connected to its equilibrium position by an ideal spring whose spring constant is k_s.

In both newtonian and quantum mechanics, we can treat a particle oscillating in three dimensions as if it were three independent *one*-dimensional oscillators. For example, the total newtonian energy of a three-dimensional harmonic oscillator is

The model treats atoms as independent oscillators

We can treat a three-dimensional oscillator as three one-dimensional oscillators

$$E = \tfrac{1}{2}mv^2 + \tfrac{1}{2}k_s r^2 = \tfrac{1}{2}m\left(v_x^2 + v_y^2 + v_z^2\right) + \tfrac{1}{2}k_s(x^2 + y^2 + z^2)$$

$$= \left(\tfrac{1}{2}mv_x^2 + \tfrac{1}{2}k_s x^2\right) + \left(\tfrac{1}{2}mv_y^2 + \tfrac{1}{2}k_s y^2\right) + \left(\tfrac{1}{2}mv_z^2 + \tfrac{1}{2}k_s z^2\right) \quad \text{(T4.1)}$$

where m is the atom's mass. Note how the terms in the energy equation can be grouped into three pairs, each pair of which would be the energy associated with a one-dimensional oscillation along one of the coordinate axes, without any reference to what is happening in the other coordinate directions. One can also show from this equation (see problem T4A.1) that the

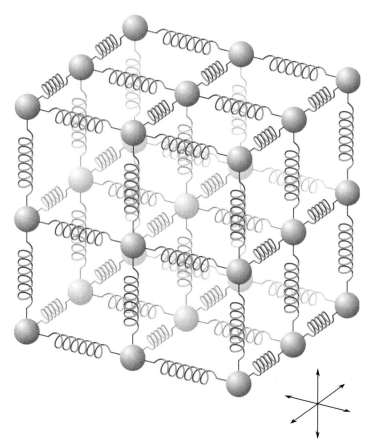

Figure T4.1

In Einstein's model of a monatomic crystalline solid, we imagine
the atoms to be held in their positions by springs.

atom's motion along each coordinate axis is exactly as if the atom were oscillating in one dimension along that axis alone. (This is a special property of
the harmonic oscillator potential energy function; most other potential energy functions cannot be pulled apart in this way.)

According to quantum mechanics, the energy associated with each of
these separate one-dimensional oscillations is quantized, so that the atom's
total vibrational energy is given by

$$E = \hbar\omega\left(n_x + \tfrac{1}{2}\right) + \hbar\omega\left(n_y + \tfrac{1}{2}\right) + \hbar\omega\left(n_z + \tfrac{1}{2}\right) \qquad \text{(T4.2)}$$

where $\omega = (k_s/m)^{1/2}$ is the angular frequency of the equivalent newtonian
oscillator and $\hbar \equiv h/2\pi$ where h is Planck's constant. The quantities n_x, n_y,
and n_z here are independent, nonnegative integers (0, 1, 2, 3, and so on) that
specify each oscillator's energy level. Each of the three terms here is the same
as the expression for the energy of a *one*-dimensional quantum harmonic oscillator (as we saw in unit Q).

We can rewrite this equation as

$$E = \sum_{i=1}^{3} \hbar\omega\left(n_i + \tfrac{1}{2}\right) = \sum_{i=1}^{3} \varepsilon\left(n_i + \tfrac{1}{2}\right) \qquad \text{(T4.3)}$$

where $\varepsilon \equiv \hbar\omega$ is the energy difference between adjacent levels of each one-
dimensional oscillator and n_1, n_2, and n_3 are just a different way of labeling

the integers n_x, n_y, and n_z. We can then find the solid's total energy by summing this over all atoms (three terms per atom):

$$E_{\text{tot}} = \sum_{i=1}^{3N} \varepsilon \left(n_i + \frac{1}{2} \right) = \sum_{i=1}^{3N} \varepsilon n_i + \sum_{i=1}^{3N} \frac{1}{2}\varepsilon = \sum_{i=1}^{3N} \varepsilon n_i + \frac{3}{2}N\varepsilon \qquad \text{(T4.4)}$$

Now, the constant $(3/2)N\varepsilon$ term in this equation is called the solid's **zero-point energy.** The solid will have this energy at absolute zero (which is the temperature where all atoms are in their *lowest possible* energy state, by definition). We usually define a system's *thermal* energy U in a given macrostate to be the difference between its energy in that state and the energy it would have at $T = 0$, so that U expresses the part of the system's energy that can *change* in thermodynamic processes. Therefore, we do not consider zero-point energy to be part of the solid's *thermal* energy U. So

The quantum zero-point energy of the solid is irrelevant

$$U = \sum_{i=1}^{3N} \varepsilon n_i \qquad \text{(T4.5)}$$

The thermal energy of an Einstein solid

Purpose: This equation specifies the total thermal energy U of an Einstein solid whose N atoms we model as $3N$ identical but independent quantum harmonic oscillators.
Symbols: n_1, n_2, n_3, . . . are a set of nonnegative integers (three per atom), and ε is the fixed difference between energy levels of each oscillator.
Limitations: The Einstein model only works well for monatomic solids at temperatures above 100 K or so (the exact limit depends on the solid).

Note that $\varepsilon \equiv \hbar(k_s/m)^{1/2}$, where k_s is the effective spring constant of the interactions holding an atom to its equilibrium position in the solid, m is the mass of the atom, and $\hbar$ is Planck's constant h over 2π. Thus ε increases as the strength of the interatomic forces increases and decreases as the atomic mass increases.

The basic point of equation T4.5 is that *we will model a crystalline solid containing N identical atoms as if it contained 3N identical independent quantum harmonic oscillators, each of which can store an integer number n_i of energy units $\varepsilon \equiv \hbar\omega$.* We will call any solid accurately described by this model an **Einstein solid.**

T4.3   Counting Microstates

From this section to the end of chapter T5, we will focus on learning what we can about the thermal behavior of this particular model for a solid. Why focus on this model, and not, say, the ideal gas model? The answer is that it is *much* easier to determine the number of microstates corresponding to each macrostate for an Einstein solid than for any other reasonably realistic model. This in turn makes it comparatively easier to determine what statistical physics predicts about this model. However, models of other complex systems (including the ideal gas model) are *qualitatively* similar to this

Why focus on Einstein solids?

model, so what we learn from close inspection of this model applies in qualitative terms at least to other systems as well (as we will discuss in chapter T6).

Describing the macrostate of an Einstein solid

To describe the macrostate of an Einstein solid, it is sufficient to specify the number of atoms N and its total internal energy U. Why? Empirically, we know that $U = 3Nk_BT$ for a monatomic solid; so if we know U, we know T. We assume in the Einstein model that each atom occupies a fixed volume independent of the applied pressure P and is independent of other atoms, so P is irrelevant and we can calculate V directly from N. (The volume of a typical *real* solid depends only very weakly on temperature and pressure, so our model is a reasonably good approximation.) Therefore, N and U describe everything we need to know about the solid at the macroscopic level. (That we need *three* variables instead of two to describe the macrostate of an ideal gas is one of the things that makes that model more complicated than the Einstein solid for our purposes here.)

Definition of *multiplicity*

To describe a microstate of the solid, we have to specify an integer value n_i for each of the solid's $3N$ independent oscillators. In general there will be *many* microstates (i.e., many distinct sets of values for all $3N$ integers n_i) that have the same total U. The number of possible microstates corresponding to the same given macrostate is called the **multiplicity** Ω of that macrostate. In the case of an Einstein solid, where the macrostate is specified by U and N,

$$\Omega(N, U) \equiv \text{multiplicity of macrostate specified by } U \text{ and } N$$
$$= \text{number of } N\text{-atom microstates having total energy } U \quad \text{(T4.6)}$$

Counting microstates

How can we determine $\Omega(U, N)$ for given values of U and N? The beauty of the Einstein solid model is that this is not a *conceptually* difficult problem. According to equation T4.5, the total energy in an Einstein solid is an integer multiple of the basic energy unit ε. Think of each energy unit as a marble, and each of the solid's $3N$ oscillators as a bin into which we can put marbles. When we specify the solid's total energy U, we are essentially specifying the total number of "marbles" $q \equiv U/\varepsilon$ that we have to distribute. Counting the microstates for this U, then, is the same as counting how many different ways we can sort q marbles into $3N$ bins.

Examples of counting microstates for small U, N

In a solid of macroscopic size, $3N$ will be on the order of magnitude of 10^{23}, but let us start small. Imagine we have an Einstein solid consisting of a *single* atom, that is, *three* independent oscillators. (Of course, a single atom will have no lattice in which it can oscillate, but let us just pretend that this makes sense. We'll work up to larger numbers of atoms shortly.) We can describe the microstate of this system by specifying energy-level integers for each of the three oscillators (i.e., how many marbles each of these three bins contains). Let us write these numbers as a triplet of digits; for example, the triplet 032 specifies the microstate in which the first oscillator has 0 units of energy, the second has 3 units, and the third has 2 units. The total energy contained in the system in this case is $U = 5\varepsilon$. Other possible microstates corresponding to this macrostate are 320, 230, 302, 203, 023, 113, 311, 131, 041, 014, etc.

So let us start counting microstates for various different macrostates of this hypothetical one-atom Einstein solid. First, imagine that the total energy of the solid has its lowest possible value $U = 0$. There is only one microstate (000) compatible with this total energy, so we say that the multiplicity of this macrostate is $\Omega(N, U) = \Omega(1, 0\varepsilon) = 1$.

Now imagine that the total energy of the solid is $U = \varepsilon$ (that is, the solid contains exactly 1 unit of energy). The microstates compatible with this total energy are 100, 010, and 001, for a total of three. The multiplicity of this macrostate is thus $\Omega(1, 1\varepsilon) = 3$.

Exercise T4X.1

Now imagine that the total energy of the solid is $U = 2\varepsilon$. The multiplicity of this macrostate turns out to be $\Omega(1, 2\varepsilon) = 6$. Write down the triplets for the six microstates corresponding to this macrostate.

Exercise T4X.2

Now imagine that the total energy of the solid is $U = 3\varepsilon$. What is the multiplicity of this macrostate? (Write down all possible microstate triplets consistent with this macrostate and count them.)

Similarly, $\Omega(1, 4\varepsilon) = 15$, $\Omega(1, 5\varepsilon) = 21$, and $\Omega(1, 6\varepsilon) = 28$.

An Einstein solid consisting of two atoms (six independent oscillators) and zero total energy has one microstate 000000, so $\Omega(2, 0\varepsilon) = 1$. If it has a total energy of $U = \varepsilon$, then the six possible microstates are 000001, 000010, 000100, 001000, 010000, and 100000, meaning that $\Omega(2, 1\varepsilon) = 6$. If it has a total energy of $U = 2\varepsilon$, then the possible microstates are 000002, 000020, 000200, 002000, 020000, and 200000; and 000011, 000101, 001001, 010001, 100001, 000110, 001010, 001100, 010010, 010100, 011000, 100010, 100100, 101000, and 110000, for a total of 21 states, so $\Omega(2, 2\varepsilon) = 21$. In a similar fashion, you can determine that $\Omega(2, 3\varepsilon) = 56$, $\Omega(2, 4\varepsilon) = 126$, and so on.

You can see that this counting of microstates gets pretty tedious after a while. There is in fact a general formula for the number of microstates for a given system. If $q = U/\varepsilon$ is the total number of energy units to be distributed among $3N$ oscillators, then the multiplicity of the macrostate is

$$\Omega(N, U) = \frac{(q + 3N - 1)!}{q!(3N - 1)!} \tag{T4.7}$$

The multiplicity of an Einstein solid's macrostate

Purpose: This equation specifies the multiplicity Ω of a macrostate of an Einstein solid.

Symbols: N is the number of atoms in the solid; U is its total energy; $q \equiv U/\varepsilon$ is the number of units of energy to be distributed among the atoms; $n!$ or **n factorial** $\equiv 1 \cdot 2 \cdot 3 \cdots (n - 1) \cdot n$.

Limitations: This equation only applies to an Einstein solid.

(The derivation of this formula is discussed in problem T4S.8.) So, for example,

$$\Omega(2, 4\varepsilon) = \frac{(4 + 6 - 1)!}{4!(6 - 1)!} = \frac{9!}{4!5!} = \frac{9 \cdot 8 \cdot 7 \cdot 6 \cdot 5 \cdot 4 \cdot 3 \cdot 2 \cdot 1}{(4 \cdot 3 \cdot 2 \cdot 1)(5 \cdot 4 \cdot 3 \cdot 2 \cdot 1)}$$

$$= \frac{9 \cdot 8 \cdot 7 \cdot 6}{4 \cdot 3 \cdot 2 \cdot 1} = 9 \cdot 7 \cdot 2 = 126 \tag{T4.8}$$

Exercise T4X.3

(a) Check that formula T4.7 yields the same results for $\Omega(1, 0)$, $\Omega(1, \varepsilon)$, $\Omega(1, 2\varepsilon)$, and $\Omega(2, 2\varepsilon)$ that we found earlier by direct counting. (b) Use formula T4.7 to verify that $\Omega(1, 6\varepsilon) = 28$. (c) If an Einstein solid has three atoms, what is the multiplicity of the macrostate where it has 8 units of energy?

T4.4   Two Einstein Solids in Thermal Contact

Suppose that we now bring two Einstein solids A and B, one with N_A atoms (three N_A oscillators) and one with N_B atoms (three N_B oscillators), into thermal contact, so that microscopic interactions between atoms on the surfaces in contact can allow energy to flow between the solids. What we would like to understand, however, is how these solids will behave *macroscopically* after being brought into contact.

The macrostate of solid A is specified by N_A and U_A, while the macrostate of solid B is specified by N_B and U_B. Since N_A and N_B are fixed, the macrostates of A and B are essentially determined by their respective energies U_A and U_B. If the combined system of the two solids is thermally isolated, its total energy $U = U_A + U_B$ is fixed (by conservation of energy); but at least in principle, the energies U_A and U_B of the two solids could have any values consistent with that total. For example, if the combined system's total energy is $U = 6\varepsilon$, then possible pairs of values for U_A and U_B include $U_A = 0$ and $U_B = 6\varepsilon$, or $U_A = 2\varepsilon$ and $U_B = 4\varepsilon$, or $U_A = 5\varepsilon$ and $U_B = \varepsilon$, and so on.

The *macropartition* of a pair of objects in thermal contact

Let us call a given pair of macrostates for solids A and B that are consistent with a fixed value of $U = U_A + U_B$ a **macropartition** of the combined system for that U. For example, the pair of macrostates where $U_A = 2\varepsilon$ and $U_B = 4\varepsilon$ is one possible *macropartition* of the combined system for $U = 6\varepsilon$.

Different macropartitions of the combined system of two solids therefore amount to different ways that the energy can be *macroscopically* divided (or "partitioned") between the solids. There is a real distinction to be made here between a *macropartition* and a *microstate* of the combined system. A microstate of the combined system specifies exactly how much energy *each individual oscillator* in both solids has. A macropartition, on the other hand, only specifies the *macroscopic* total energies U_A and U_B that the two macroscopic solids have, something we can measure macroscopically. In other words, we describe a macropartition of a combined system of two subsystems by describing *the macrostate of each subsystem*.

A macropartition table for Einstein solids in contact

It is easiest to illustrate these ideas with a specific example. Suppose we bring two hypothetical one-atom solids into thermal contact. Imagine that their total combined energy is $U = 6\varepsilon$. The **macropartition table** T4.1 lists all possible macropartitions for a combined system with this total energy.

Table T4.1   Possible macropartitions for $N_A = N_B = 1$, $U = 6\varepsilon$

Macropartition	U_A	U_B	Ω_A	Ω_B	Ω_{AB}
0:6	0	6	1	28	28
1:5	1	5	3	21	63
2:4	2	4	6	15	90
3:3	3	3	10	10	100
4:2	4	2	15	6	90
5:1	5	1	21	3	63
6:0	6	0	28	1	28

Grand total number of microstates $= 462$

The numbers under the columns labeled U_A and U_B specify the macrostates of solids A and B by specifying the total energy of each solid in units of the basic energy unit ε. Each number under the heading Ω_A states the multiplicity of solid A when it has the specified energy U_A, and each number under the heading Ω_B states the same for solid B. The value of Ω_{AB} specified for each macropartition is the total multiplicity of the combined system in that macropartition (i.e., the total number of distinct microstates available to the system as a whole in that macropartition). This total multiplicity Ω_{AB} is the product of Ω_A and Ω_B

$$\Omega_{AB} = \Omega_A \Omega_B \qquad \text{(T4.9)}$$

because for each one of the Ω_A possible microstates for solid A, there are Ω_B possible microstates for solid B. For example, in the macropartition 3:3, the possible microstates of solid A are (in our previous notation) 300, 030, 003, 210, 201, 021, 120, 102, 012, and 111, and the possible microstates of system B are the same. The possible microstates of the combined system are as follows (the triplets on the left and right specify the microstates of solids A and B, respectively): 300-300, 300-030, 300-003, 300-210, 300-201, 300-021, 300-120, 300-102, 300-012, 300-111, 030-300, 030-030, 030-003, 030-210, and so on, for a total of $10 \times 10 = 100$ distinct microstates.

Exercise T4X.4

Prepare an analogous table for the case where $N_A = N_B = 1$ and $U = 8\varepsilon$. (Most of the multiplicities Ω_A and Ω_B you can copy from table T4.1; use equation T4.7 to calculate the rest.)

T4.5   The Fundamental Assumption

In a more familiar system such as the ideal gas model, it is obvious that the gas will continually and randomly shift from microstate to microstate as the molecules collide with the walls and one another and exchange energy. A similar situation will apply here if we assume that the elementary oscillators in each solid weakly interact with one another: not so strongly that the approximation that the oscillators are independent becomes invalid, but strongly enough that energy is freely interchanged between the oscillators. Energy will also be shifted randomly back and forth across the boundary between the solids through the interactions of the oscillators on the surfaces in contact.

In short, as time passes, the combined system of two solids will randomly shift between different microstates consistent with the constraint that the total energy have some fixed value U. This means that under some circumstances, the macropartition of the combined system might fluctuate as the system randomly samples microstates in different macropartitions. For example, in the situation considered in table T4.1, the combined system in microstate 012-300 (one of the microstates corresponding to macropartition 3:3) might evolve to 013-200 (one of the microstates corresponding to macropartition 4:2) by transfer of 1 unit of energy across the boundary. In time, this system will sample each of the 462 possible microstates, and thus each of the possible macropartitions.

Now comes the big question: Can we say something about which of the macropartitions are the ones that we are most likely to see if we peek at the system at various times? The answer is yes, if we are willing to accept a simple and plausible assumption about the behavior of such systems. This assumption is called the **fundamental assumption** of statistical mechanics:

All of a system's accessible microstates are equally likely in the long run.

Accessible in this context means "consistent with the value of the total internal energy of the system in question."

This disarmingly simple postulate provides the foundation for understanding irreversible processes, as we will see in chapters T5 and T6. Note that even though this assumption is simple and plausible, its ultimate justification is that it correctly predicts the behavior of macroscopic systems.

The most important consequence of this principle for us is that the probability of occurrence of a given energy macropartition (consistent with the given total internal energy) is directly proportional to the number of microstates that indistinguishably generate that macropartition, that is, to the total multiplicity Ω_{AB} of that macropartition.

For example, suppose that we were to take a large number of "snapshots" of the system of two Einstein solids described in table T4.1. The fundamental assumption means that we would find the system to be in macropartition 3:3 in about $100/462 = 0.216 = 21.6\%$ of the pictures and in macropartition 0:6 (or macropartition 6:0) in about $28/462 = 0.061 = 6.1\%$ of the pictures and so on. Note that macropartition 3:3, the macropartition for which the energy is shared equally between the two identical solids, is the single most probable macropartition of the system.

Exercise T4X.5

(a) Compute the probabilities of each of the macropartitions in table T4.1, and write them to the right of the value of Ω_{AB} for that macropartition. (b) Do the same for the table that you constructed in exercise T4X.4. (c) Which macropartition in the latter case is the most probable one?

T4.6   Using StatMech

Doing the calculations required to set up a table such as table T4.1 can be very tedious, particularly as the number of atoms in each solid becomes large. Fortunately, you can download a free computer program called StatMech that does all the calculations for you. This program is available in both Macintosh and Windows versions; go to the *Six Ideas* website (whose URL is given in this book's preface) and follow the links to the computer programs page.

When you run the program, it displays a window that looks something like that shown in figure T4.2 (except without the table). In the text boxes at the top of the window, you can enter the number of atoms you want in solids A and B and the total energy U that these solids will share (expressed as a multiple of the basic energy unit ε). If you then press the Calculate button, the program displays a table in the lower part of the window, shown in

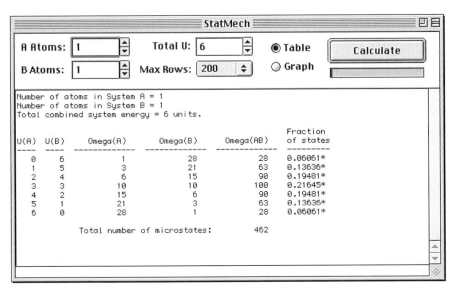

Figure T4.2
A screen shot of StatMech set up to display the macropartition table shown in table T4.1.

figure T4.2. You can print this table using the Print command under the File menu, or you can select all or part of the table and paste it into a word processing program.

Note that the final column of the table lists the number of microstates for each macrostate expressed as a fraction of the total number of accessible microstates. According to the fundamental assumption of statistical mechanics, this is also the probability of that macrostate's occurring. The starred macropartitions have probabilities above 0.0001; this means that the starred macrostates embrace about 99.98% of the system's total microstates.

In principle, you can do by hand what StatMech does, at least when N_A, N_B, and U are reasonably small (indeed, you did this in exercises T4X.4 and T4X.5). Once you have constructed a few tables on your own, however, you will appreciate how rapidly and easily the program generates tables.

Once you have calculated a table, you can also display a graph of the macropartition probabilities as a function of U_A by pressing the Graph radio button. The graph corresponding to the table in figure T4.2 is shown in figure T4.3.

You can enter fairly large numbers for N_A and N_B without any problem. If you enter numbers for U that are larger than the Max Rows (maximum number of table rows) value selected, each row of the table will display the summed results for a set of macropartitions instead of the results for an individual macropartition. The program does this so that the number of table rows does not become too unwieldy.

How StatMech handles large values of U

The table layout becomes somewhat different under these circumstances: Instead of columns for U_A and U_B, you will see columns specifying the set number (equivalent to the table row number) and the average values of the ratios U_A/U and U_B/U for the macropartitions in the set. An example is shown in figure T4.4. Under these circumstances, the program will only accept a value of U that is an integer multiple of Max Rows minus 1, so that the number of macropartitions in every bin is exactly the same. If you enter a number that is just an integer multiple of Max Rows instead of a multiple minus 1 (as I did in the case shown in figure T4.4), the program

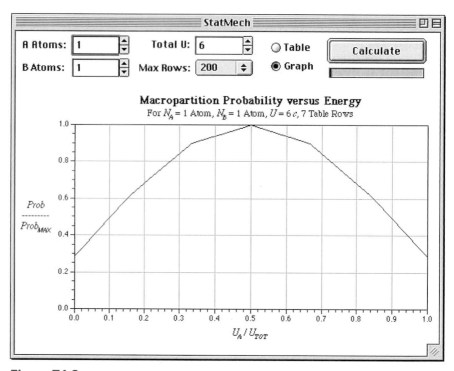

Figure T4.3
A graph of the relative macropartition probabilities for the table shown in figure T4.2.

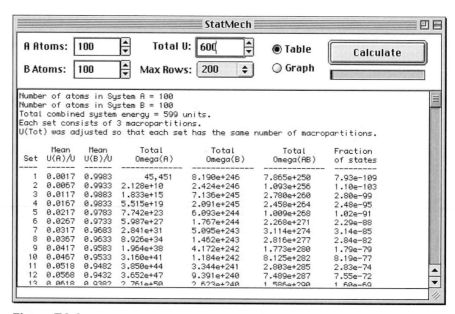

Figure T4.4
This is how StatMech formats the macropartition table when the value for *U* exceeds Max Rows.

assumes that you really meant the number typed minus 1. (Specifying a value for U that is bigger than Max Rows but is not an integer multiple of Max Rows or an integer multiple minus 1 generates an error message when you press Calculate.)

For more complete (and/or up-to-date) instructions, select About StatMech (under the Apple menu on Macintosh systems or under the Edit menu on Windows systems).

Before you read chapter T5, I strongly recommend that you play with this program. First check that the program reproduces the results of exercises T4X.4 and T4X.5. Then try larger solids with larger total energies. Do some runs where the two solids have unequal sizes as well. This program plainly displays Boltzmann's resolution to the mystery of irreversibility, if you have the eyes to see it! See if *you* can figure it out *before* you read chapter T5. The truth is out there.

The truth about irreversibility is out there . . .

[*Hint:* Look at the table for a set of two solids with fairly large values of N_A, N_B, and U. Imagine that we start out with solid A being very cold (U_A very small) and solid B being very hot ($U_B \approx U$). What will happen as time passes and the pair of solids starts visiting microstates in adjacent macropartitions?]

Exercise T4X.6

(a) Run StatMech for the case where $N_A = N_B = 1$ and $U = 6\varepsilon$, and verify that the program reproduces table T4.1. Also have a look at the graph of macropartition probabilities. (b) Run the case where $N_A = N_B = 10$ and $U = 60\varepsilon$. What is different about the table (besides the fact that it is longer)? What is different about the graph? (c) Now scale everything up by a factor of 10 again, to $N_A = N_B = 100$ and $U = 600\varepsilon$. (Note that the program now begins to arrange macropartitions into bins.) What else has changed about the table? How has the graph changed? (d) Continue to increase the size of the system in steps of factors of 10 until the calculations start to take too long. Describe what you think are the important trends in terms of how the tables and graphs change as the size of the system increases.

TWO-MINUTE PROBLEMS

T4T.1 Consider a system consisting of two Einstein solids P and Q in thermal contact. Assume that we know the number of atoms in each solid and ε. What do we know about the system if we also know the quantum state of each atom in each solid?
A. Its macrostate
B. Its microstate
C. Its macropartition
D. Its microstate and macropartition
E. Its macrostate and macropartition
F. Its macrostate and microstate
T. Its macrostate, microstate, and macropartition

T4T.2 Consider a system consisting of two Einstein solids P and Q in thermal contact. Assume that we know the number of atoms in each solid and ε. What do we know about the system if we also know the total energy in each of the two objects? (Please choose from the possible answers listed in problem T4T.1.)

T4T.3 Consider a system consisting of two Einstein solids P and Q in thermal contact. Assume that we know the number of atoms in each solid and ε. What do we know about the system if we also know the total energy of the combined system? (Please choose from the possible answers listed in problem T4T.1.)

T4T.4 What is the *crucial* characteristic of an Einstein solid that makes it easier to analyze in the context of this chapter than most other kinds of thermodynamic systems?
A. The atoms are arranged in a regular, cubic lattice.
B. The atoms are identical.
C. Its microstates are comparatively easy to count.
D. The atomic vibration energy levels are equally spaced.
E. $\Omega(U, N)$ is always reasonably small.
F. Other (specify).

T4T.5 The zero-point energy of an Einstein solid can be ignored because
A. It is zero.
B. It never changes in any thermal interaction.
C. It is insignificant compared to the solid's total energy.

D. It is just a quantum-mechanical effect.
E. Other (specify).

T4T.6 The multiplicity of an Einstein solid with 3 atoms and 4 units of energy is
A. 4.8×10^8
B. 715
C. 495
D. 36
E. 12
F. Other

T4T.7 Which of the following statements is true?
A. There are always many microstates in a macrostate.
B. All accessible macrostates are equally probable.
C. All microstates of a system are equally probable.
D. All accessible macropartitions are equally probable.
E. When two objects in thermal contact are isolated from everything else, their macrostates cannot change.
F. None of the above.

HOMEWORK PROBLEMS

Basic Skills

T4B.1 Imagine that we have an ideal gas consisting of 15 molecules. We can flip the signs of each of the three velocity components of a given molecule without changing its overall energy (and thus without changing the gas's macrostate). How many possible patterns of sign choices are there?

T4B.2 Calculate the multiplicity of an Einstein solid with $N = 1$ and $U = 6\varepsilon$ by directly listing and counting the microstates. Check your work by using equation T4.7.

T4B.3 Calculate the multiplicity of an Einstein solid with $N = 1$ and $U = 5\varepsilon$ by directly listing and counting the microstates. Check your work by using equation T4.7.

T4B.4 Use equation T4.7 to calculate the multiplicity of an Einstein solid with $N = 4$ and $U = 10\varepsilon$.

T4B.5 Use equation T4.7 to calculate the multiplicity of an Einstein solid with $N = 3$ and $U = 15\varepsilon$.

T4B.6 How many times more likely is it that the combined system of solids described in table T4.1 will be found in macropartition 3:3 than in macropartition 0:6, if the fundamental assumption is true?

T4B.7 How many times more likely is it that the combined system of solids described in table T4.1 will *not* be

found in macropartition 3:3 than it is to be found in macropartition 0:6, if the fundamental assumption is true?

Synthetic

T4S.1 Consider an Einstein solid consisting of $N_A = 1$ atom (three oscillators). Each oscillator can store any integer number of energy units ε. The following table lists the number of microstates Ω_A available to the solid when it has various values of total thermal energy U_A.

U_A	0	1ε	2ε	3ε	4ε	5ε	6ε	7ε	8ε	9ε
Ω_A	1	3	6	10	15	21	28	36	45	55

By actually listing and counting the various possible microstates, verify the results for the multiplicity Ω_A for the cases where $U_A = 4\varepsilon$ and $U_A = 7\varepsilon$. (You can check your results by using equation T4.7.)

T4S.2 Consider an Einstein solid consisting of $N_B = 2$ atoms. The following table lists the number of microstates Ω_B available to the solid when it has various values of total internal energy U_B.

U_B	0	1ε	2ε	3ε	4ε	5ε	6ε	7ε	8ε	9ε
Ω_B	1	6	21	56	126	252	462	792	1287	2002

(a) By actually listing the various possible microstates, verify the result for $U_B = 2\varepsilon$.

(b) Using equation T4.7, verify the value of Ω_B for $U_B = 6\varepsilon$ and 9ε.

T4S.3 Imagine putting the two solids discussed in problems T4S.1 and T4S.2 into thermal contact. Imagine that the resulting combined system is isolated from everything else, and that the combined system contains 6 units of energy (that is, $U_A + U_B = 6\varepsilon$).

(a) Construct a table showing U_A, U_B, Ω_A, Ω_B, and Ω_{AB} for all possible macropartitions of the system (i.e., a table analogous to table T4.1), and compute the probabilities for each of the seven possible macropartitions according to the fundamental assumption.

(b) Identify the most probable macropartition(s) of this system. Is the energy evenly divided between the solids in the most probable macropartition(s)?

T4S.4 Imagine putting the two solids discussed in problems T4S.1 and T4S.2 into thermal contact. Imagine that the resulting combined system is isolated from everything else, and that the combined system contains 9 units of energy (that is, $U_A + U_B = 9\varepsilon$).

(a) Construct a table showing U_A, U_B, Ω_A, Ω_B, and Ω_{AB} for all possible macropartitions of the system (i.e., a table analogous to table T4.1), and compute the probabilities for each of the 10 possible macropartitions according to the fundamental assumption.

(b) Identify the most probable macropartition(s) of this system. Is the energy evenly divided between the solids in the most probable macropartition(s)?

T4S.5 Imagine putting two solids with $N_A = N_B = 2$ in thermal contact, and imagine that the resulting combined system is isolated from everything else and that it contains 9 units of energy (that is, $U_A + U_B = 9\varepsilon$).

(a) Construct a table showing U_A, U_B, Ω_A, Ω_B, and Ω_{AB} for all possible macropartitions of the system (i.e., a table analogous to table T4.1), and compute the probabilities for each of the 10 possible macropartitions according to the fundamental assumption.

(b) Identify the most probable macropartition(s) of this system. Is the energy evenly divided between the solids in the most probable macropartition(s)?

T4S.6 (a) Use StatMech to generate a macropartition table for the situation where $N_A = N_B = 5$ and $U = 20\varepsilon$, and answer the following questions. (1) How many total microstates are available to the system? (2) Which is the most probable macropartition, and how many microstates are available to the system in this macropartition? (3) What is the average energy per atom in each solid in this macropartition? (4) What range of values of the ratio U_A/U corresponds to macropartitions

whose probabilities are at least one-half as large as the most probable macropartition? (*Hint:* You can see this pretty quickly on the graph.)

(b) Answer the same set of questions as in (a) for the case where $N_A = N_B = 50$ and $U = 200\varepsilon$ and for the case where $N_A = N_B = 500$ and $U = 2000\varepsilon$. (Note that I am scaling all variables up by a factor of 10 each time, so that the energy per atom remains fixed.)

(c) Carefully and completely describe any trends you see in how the answers to these questions change as the system becomes larger.

T4S.7 (a) Use StatMech to generate tables for the following five different cases: (1) $N_A = N_B = 50$, (2) $N_A = 40$ and $N_B = 60$, (3) $N_A = 30$ and $N_B = 70$, (4) $N_A = 20$ and $N_B = 80$, and (5) $N_A = 10$ and $N_B = 90$. Choose $U = 200\varepsilon$ in all cases. In each case, compare the energy per atom for each solid (i.e., compare U_A/N_A to U_B/N_B) when the system is in the most probable macropartition. (You can see the big picture pretty quickly if you look at the graphs of macropartition probabilities; look at the table for details.) State a simple rule describing how U_A/N_A is related to U_B/N_B in the most probable macropartition. Is your rule always exactly true?

(b) How does your rule fare when you increase all numbers by a factor of 100? Support your response with some evidence.

T4S.8 We can derive equation T4.7 as follows. First, note that the problem of counting the microstates of an Einstein solid is essentially the same as the problem of finding the number of distinct patterns that can be generated by pulling marbles and matchsticks randomly from a bag. For example, imagine that we pull the following sequence of items from the bag (reading from left to right):

If we imagine each marble to be a unit of energy and each matchstick to be a *division* between two oscillators, then this pattern corresponds to the 130211 microstate for an Einstein solid with $N = 2$ atoms (six oscillators) and $U = 8\varepsilon$.

(a) Argue that a solid with M oscillators and q units of energy will be represented by q marbles and $M - 1$ matches.

(b) Argue that if we put $M - 1$ matches and n marbles in the bag, there will be a total of $(M + q - 1)!$ different ways of pulling objects out of the bag. (*Hint:* When we select the first object, we have $M + q - 1$ choices. When we choose the second item, we now only have $M + q - 2$ choices, since we've already pulled out one object, and so on.)

(c) Not all these distinct ways of pulling out objects generate distinct patterns. For example, consider taking a given pattern and rearranging the marbles. The rearrangement does not change the basic pattern, but would represent a different sequence of choices as we pull objects out of the bag, and thus would be counted as a distinct choice in (b). Argue that there are $q!$ ways of rearranging the marbles and $(M-1)!$ ways of rearranging the matchsticks without affecting the pattern.

(d) Argue finally that equation T4.7 correctly states the number of distinct patterns that can be generated, and thus the number of distinct microstates of an Einstein solid with M oscillators and n units of energy.

Rich-Context

T4R.1 Consider a hypothetical system in which each atom can only be in one of two quantum states, a ground state with energy 0 or an excited state with energy ε. The atoms have fixed positions and fixed volumes, and there is no way for an atom to change its energy except by going from the ground state to the excited state or vice versa. As with the Einstein solid, we can completely describe the macrostate of this system by specifying the number of atoms N and their total thermal energy U.

(a) Find an expression, in terms of U and N, for the number of atoms n that are in the excited state.

(b) Find a formula for the multiplicity $\Omega(U, N)$ of a macrostate where the system has N atoms and total energy U. (*Hints:* Play around with small systems first. For example, imagine that we have $N = 3$ and $U = 2\varepsilon$. We might describe the possible microstates as being 011, 101, and 110. Once you have developed some sense about the multiplicity of small systems, start working on a general formula, which will involve some factorials. Once you have a trial formula, go back and check it in small-system cases where you can directly count the microstates.)

(c) Construct a macropartition table for a system consisting of two subsystems with $N_A = 20$, $N_B = 20$, and $U = 8\varepsilon$. Does it look at least qualitatively like the macropartition tables we have been constructing for Einstein solids? Explain.

Comment: This model is also actually a pretty good description of certain kinds of solids at very low temperatures when they are placed in a strong magnetic field. Under such circumstances, the magnetic moment of an atom in the solid can be either aligned with the field (which we can take to be the ground state) or anti-aligned with the field (which we can take to be the excited state). It is also easier to determine multiplicities for this model than for the Einstein solid. So why do we not focus on this model instead of on the Einstein solid? This model only applies to pretty esoteric systems. Even worse, it exhibits peculiar behavior when its energy is such that $n > \frac{1}{2}N$, behavior that is almost never seen in nature otherwise. Therefore, unlike the Einstein solid, this model is not very useful for helping us understand how normal complex systems behave.

Advanced

T4A.1 (a) Show that if you take the time derivative of equation T4.1, you get

$$0 = v_x(ma_x + k_s x) + v_y(ma_y + k_s y) + v_z(ma_z + k_s z)$$

$$\text{(T4.10)}$$

(b) By changing initial conditions, I could arrange it so that v_x, v_y, and v_z have any values I please at a given time. Argue that since this equation has to be zero at *all* times, the quantities in parentheses have to be *independently* equal to zero at all times. (c) Show that each of these quantities in parentheses is the same as Newton's second law for a simple one-dimensional harmonic oscillator moving in the corresponding axis direction.

ANSWERS TO EXERCISES

T4X.1   011, 101, 110, 200, 020, 002

T4X.2   111, 012, 021, 102, 201, 120, 210, 003, 030, and 300, so $\Omega(1, 3\varepsilon) = 10$

T4X.3 $\Omega(3, 8\varepsilon) = 12{,}870$

T4X.4   The table is as follows:

Macropartition	U_A	U_B	Ω_A	Ω_B	Ω_{AB}
0:8	0	8	1	45	45
1:7	1	7	3	36	108
2:6	2	6	6	28	168
3:5	3	5	10	21	210
4:4	4	4	15	15	225
5:3	5	3	21	10	210
6:2	6	2	28	6	168
7:1	7	1	36	3	108
8:0	8	0	45	1	45

Grand total number of microstates = 1287

T4X.5 The probabilities are as follows:

Table T4.1	
Macro-partition	Probability
0:6	0.061
1:5	0.136
2:4	0.195
3:3	0.216
4:2	0.195
5:1	0.136
6:0	0.061

Exercise T4X.4	
Macro-partition	Probability
0:8	0.035
1:7	0.084
2:6	0.131
3:5	0.163
4:4	0.175
5:3	0.163
6:2	0.131
7:1	0.084
8:0	0.035

The 4:4 macropartition is the most probable one in the list for the situation in exercise T4X.4.

T4X.6 You should find that as the systems get larger, (1) both the multiplicity associated with each row of the table and the total number of microstates available to the system become *extremely* large, (2) the most probable macropartition is always where the energy is split evenly between the two systems, but (3) the width of the peak on the macropartition probability graph becomes smaller and smaller as the system size increases. This means that for large systems, the vast majority of microstates are within macropartitions very close to the most probable macropartition.

T5

The Second Law

Chapter Overview

Introduction

This is the core chapter of the unit. In this chapter, we will develop further the ideas introduced in chapter T4, using StatMech to create macropartitions for larger and larger Einstein solids. In doing so, we will finally discover how random microscopic processes lead to irreversible behavior in macroscopic systems. We will also learn how a system's *entropy* conveniently characterizes and quantifies its irreversible behavior.

Section T5.1: What Happens as Systems Become Large?

Using StatMech, we can easily create macropartition tables for large Einstein solids sharing many units of energy. We find that as the numbers become large, (1) the number of microstates available to the system becomes *extremely* (almost incomprehensibly) huge, but (2) the vast majority of microstates are in an increasingly narrow range of macropartitions around the most probable macropartition, so that a graph of macropartition probability versus energy in either solid becomes a very narrow spike.

Section T5.2: Irreversibility in Einstein Solids

This implies that (1) if the combined system is *not* very near the most probable (equilibrium) macropartition initially, random energy transfers will inevitably move it toward that macropartition; and (2) it will subsequently *stay* very close to that macropartition (exhibiting only tiny **fluctuations** away from it), in spite of the random energy transfers between the two solids. This is so because *that is where the microstates are*. The number of microstates available to the system increases so incredibly rapidly as the system moves toward equilibrium that while it is possible *in principle* for a random energy transfer to move the system away from equilibrium, it is so unbelievably unlikely as to be impossible for all practical purposes. (This section uses a number of StatMech-generated quantitative examples to drive home the point.) Therefore, microscopically random energy transfers lead quite naturally to essentially irreversible shifts in the system's macroscopic properties!

Section T5.3: Irreversibility in General

We have been using Einstein solids as our core example because it is easy to determine the multiplicity of a combined system of two such solids. But the argument does not depend on the exact values of the multiplicity: it applies to *any* system of objects whose multiplicities are extremely rapidly increasing functions of their thermal energies. This applies to the vast majority of physical objects (indeed, I chose to discuss Einstein solids precisely because their behavior is so typical). Understanding how irreversibility arises in Einstein solids therefore gives us at least qualitative insight into how virtually all other objects behave.

Section T5.4: The Definition of Entropy

We define a system's **entropy** S in a given macrostate to be

$$S \text{ (of macrostate)} \equiv k_B \ln \Omega \text{ (of macrostate)} \qquad \text{(T5.5)}$$

Purpose: This equation defines the entropy S of a thermodynamic system in a given macrostate.
Symbols: Ω is the macrostate's multiplicity; k_B is Boltzmann's constant.
Limitations: There are none; this is a definition. However, it is difficult to determine Ω exactly in many situations.

The entropy S is simply a different way to specify a system's multiplicity Ω, but it has several advantages over specifying the multiplicity directly: (1) the logarithm converts extremely large multiplicities to more manageably sized entropies; and (2) when we put two systems in contact, the total multiplicity is the *product* of the two systems' individual multiplicities, but the total entropy is the more convenient *sum* of the individual entropies:

$$S_{AB} = S_A + S_B \qquad \text{(T5.7)}$$

Purpose: This equation describes the total entropy S_{AB} for a given macropartition of a system consisting of two subsystems A and B in thermal contact.
Symbols: S_A and S_B are the subsystems' entropies in that macropartition.
Limitations: There are none. This directly follows from the definitions of entropy and multiplicity.

Section T5.5: The Second Law of Thermodynamics

So we see that random energy transfers between large systems in thermal contact essentially *never* move the combined system to a macropartition with a smaller multiplicity (and thus smaller total entropy). The **second law of thermodynamics** expresses this truth in a short, simple phrase: *The entropy of an isolated system never decreases.* This is one of the most important and useful laws in physics. However, one should always remember that this law (as we have seen) is a consequence of more fundamental principles (that all microstates are equally probable and that the multiplicities of most macroscopic objects increase incredibly rapidly with increasing energy).

Section T5.6: Entropy and Disorder

Equation T5.7 links entropy to *multiplicity*. Popular treatments of entropy often link it to *disorder*. This often works because macrostates we consider to be "disordered" *usually* have greater multiplicities than states we consider "ordered" (e.g., there are many more ways for your room to be disordered than ordered). But this linkage can also be misleading: Sometimes macrostates that appear more orderly than others actually have more microstates (e.g., a glass filled with disorderly crushed ice has far *fewer* microstates than a glass containing the same mass of orderly appearing water). When in doubt, you should always remember that entropy is most fundamentally linked to *multiplicity*, not disorder.

T5.1   What Happens as Systems Become Large?

A quick review

In chapter T4 we developed the basic concepts of the field of physics known as *statistical mechanics*. Using the Einstein solid as the guiding example, we learned about the crucial distinction between *macrostates* and *microstates*. We learned that the multiplicity Ω of a macrostate (the number of microstates consistent with that macrostate) can be a very large number. We learned to describe the *macropartitions* of a combined system by specifying the macrostates of its subsystems. We also saw that the fundamental assumption of statistical mechanics (that all accessible microstates are equally probable) implies that when energy is shuffled around randomly in a combined system, macropartitions with larger multiplicities will be more probable.

These ideas provide the key to understanding irreversibility and the idea of entropy. The purpose of this chapter is to work through the consequences of the fundamental assumption (still in the context of Einstein solids) to discover how irreversible processes arise naturally out of the random shuffling of microstates.

StatMech enables us to study larger systems

It turns out that the irreversible nature of macroscopic processes depends crucially on the fact that an incredibly huge number of atoms are involved in any process where macroscopic objects interact. In chapter T4, we mostly considered interacting Einstein solids consisting of a few atoms sharing only a handful of units of energy, because only for small systems is it practical to do the calculations by hand. Yet even the smallest dust mote contains thousands of trillions of atoms.

The StatMech program introduced at the end of chapter T4 allows us to explore the behavior of larger solids than we can calculate by hand, although the number of atoms that the program can handle is still far smaller than any real solid would contain. Still, by running the program with an increasing number of atoms involved in each solid, we can spot trends that allow us to extrapolate to even larger numbers.

Larger systems have huge multiplicities

If you did the runs suggested in exercise T4X.6 or problem T4S.6 or T4S.7, you should have noted several things. First, you probably noticed that as N_A, N_B, and U become large, the number of accessible microstates goes *through the roof*. For example, consider a system consisting of two Einstein solids with $N_A = N_B = 64$ (so that each solid is a tiny cube only 4 atoms on a side). When the total energy is $U = 200\varepsilon$, StatMech tells us that there are a total of about 2×10^{161} microstates available to the system. Since there are very roughly 10^{80} protons, neutrons, and electrons in the visible universe, if each proton, neutron, and electron in our universe itself contained a universe's worth of smaller particles, the total number of particles would *still* be more than a factor of 20 smaller than the total number of microstates here!

The numbers only get more ridiculous as the systems grow in size. When $N_A = N_B = 1,000,000$ (so that the solids are 100 atoms on a side) and $U = 4,999,999\varepsilon$, StatMech tells us that the total number of microstates has increased to more than $6 \times 10^{3,291,557}$. Words cannot easily express how huge this number is: it would take about 750 single-spaced pages just to *write* this number without using scientific notation. Imagine what happens as we go to solids containing roughly 10^{23} particles!

Larger systems have more sharply peaked probability graphs

Second, you may have noticed that the *shape* of the graph of macropartition probabilities changes as N_A, N_B, and U become large. When N_A, N_B, and U are small, the graph of probability versus macropartition is a fairly broad, bell-shaped curve, meaning that even the lowest-probability macropartitions have a significant probability of being seen if you watch the combined system long enough. As N_A, N_B, and U get larger, though, the curve gets narrower.

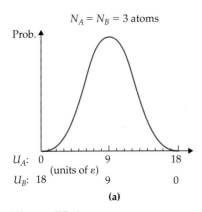
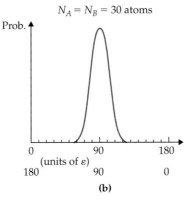
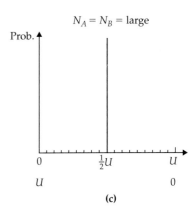

Figure T5.1
How the graph of macropartition probability changes as a system gets larger.

Perhaps you can imagine that as these numbers get very large, the probability graph narrows to a spike of almost infinitesimal width (see figure T5.1).

T5.2   Irreversibility in Einstein Solids

I claim that these results have two direct and very important implications:

1. If the combined system is *not* in the most probable macropartition to begin with, it will rapidly and inevitably move toward that macropartition.
2. It will subsequently *stay* at that macropartition (or very near to it), in spite of the random shuffling of energy back and forth between the two solids.

How do these statements follow from what we have observed?

Let us consider the second implication first. Imagine that our two Einstein solids have nearly the same energy before they are brought into contact. After they are brought into contact, they randomly shuffle energy around internally and across the boundary between them, sampling various possible microstates. The fundamental assumption implies that the probability that the system will end up in a given microstate after one of these shuffles is the same for all microstates, and since some microstates are in different macropartitions, the combined system's macropartition will **fluctuate** randomly in time.

With a very small system (say, $N_A = N_B = 5$ atoms and $U = 20\varepsilon$), these fluctuations can be pretty significant. Even if the system starts out near the center macropartition, there is about one chance in 10,000 that if we peek at the system at some later time, we will find it to have wandered into either one of the extreme macropartitions (where all the energy is in one solid and none in the other). We will see fluctuations of $\pm 4\varepsilon$ (which corresponds to almost $\pm 40\%$ of the most probable value 10ε of the energy in each solid) quite commonly.

On the other hand, even in only a modestly larger system (e.g., with $N_A = N_B = 50$ atoms and $U = 200\varepsilon$), the probability graph narrows significantly, and the probability of observing large fluctuations gets much smaller. For example, the probability of observing this system to be in its most extreme observable macropartition (where $U_A = 0$ and $U_B = 200\varepsilon$) turns out to be

$$\Pr(U_A = 0) = \frac{\Omega_{AB}(U_A = 0)}{\text{grand total } \Omega_{AB}} \approx 3.88 \times 10^{-43} \qquad \text{(T5.1)}$$

Fluctuations around
equilibrium are very small

This is an extremely small number that is really zero in any practical sense (see exercise T5X.2). Fluctuations in the values of U_A and U_B will essentially be confined to a range within about 17% of the most probable energy of each solid. If we increase the size of the system by a factor of 100 to $N_A = N_B = 5000$ atoms and $U = 19,999\varepsilon$, we find that more than 99.99% of the available microstates are in macropartitions where the objects' energies are within $\pm 1.7\%$ of their most probable values. As the size of the system increases, the range of possible fluctuations in U_A and U_B (expressed as a fraction of their most probable values) continues to decrease.

So, if two solids are initially at or near the central macropartition, then they will not stray very far from that macropartition. The fundamental reason for this is that *the vast majority of microstates are in macropartitions close to the most probable one*. If each microstate is really equally probable, the system will seem to strongly prefer being in macropartitions *very* close to the most probable macropartition.

Exercise T5X.1

Run StatMech and verify that the probability quoted in equation T5.1 is roughly correct. What is the probability that the system will be found in the 1:199 macropartition?

Exercise T5X.2

The age of the universe is probably about 1.4×10^{10} y, and there are about 3×10^7 s in 1 y. If you had been peeking at a system of two Einstein solids with $N_A = N_B = 50$ atoms and $U = 200\varepsilon$ at 1 billion times per second from the day the universe began, roughly what would be the probability that you would have seen the system in its most extreme observable macropartition (0:200) by now? (The point is that even though it is *theoretically* possible for this macropartition to occur, it is *impossible* in any practical sense.)

The inevitable march to equilibrium

Now imagine that we start with the solids in a fairly extreme macropartition. Once the solids are brought into thermal contact, they will begin to randomly shuffle energy back and forth, sampling new microstates. But the vast majority of microstates near the original macropartition are to be found in the direction toward the most probable macropartition, so the macropartition of the system will almost certainly move in that direction.

For example, consider the StatMech table shown in figure T5.2. Imagine in the situation shown (where $N_A = N_B = 250$ atoms and $U = 1000\varepsilon$) that $U_A = 12\varepsilon = 0.012U$ and $U_B = 988\varepsilon = 0.988U$ when the solids are brought into contact at $t = 0$. This would correspond to solid A having a relatively low temperature initially and solid B having a relatively high temperature. Now, it takes some time for energy to flow by random processes from one solid into the other, so not all the system's theoretically possible microstates are actually "available" to the system at the instant when we put the two solids in contact. Let us say that we wait for whatever time interval is typically required for the total energy of the solid to shift by $0.001U$ (= one energy unit ε in this case) and then peek at the system again. What is the probability that we find that the "hot" solid B gave the unit of energy to the "cold" solid A rather than the other way round?

We can estimate this by making the crude assumption that *all* the microstates in the original macropartition and in the two adjacent macropartitions are equally probable and *no* microstates in *other* macropartitions are

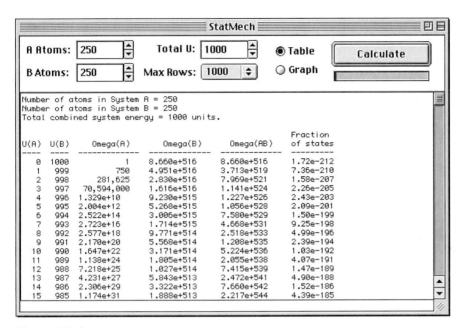

Figure T5.2
A macropartition table for a pair of solids with 250 atoms each.

possible at all. According to the table, the multiplicity of the macropartition below the initial one (the macropartition that results if B gives a unit of energy to A) is about 2.5×10^{541}. The multiplicity of the macropartition above the initial macropartition (the one that results if A gives a unit of energy to B) is about 2.1×10^{538}. This means that the probability that the hot solid B gains a unit of energy from the cold solid A compared to the probability of the reverse process is very roughly

$$\frac{\Pr(A \to B)}{\Pr(B \to A)} \approx \frac{\Omega_{AB}(U_A = 11\varepsilon)}{\Omega_{AB}(U_A = 13\varepsilon)} = \frac{2.1 \times 10^{538}}{2.5 \times 10^{541}} = 0.75 \times 10^{-30} \quad \text{(T5.2)}$$

(Note that my calculator cannot handle exponents this large, so I calculated the ratio by dividing 2.1 by 2.5 to get 0.75 and then subtracted 541 from 538 to find the exponent.) So if the energies of solids A and B change by 1 unit, it is about 1200 times more likely that the hot one will give a unit of energy to the cold one (taking the system closer to equilibrium) than it is for that energy to flow the other way.

Again, as the size of the system grows, these numbers grow more extreme. Let us scale the previous situation up by a factor of 10 (see figure T5.3). The row of the table closest to our previous initial condition is the row summarizing the 10 macropartitions whose average value of U_A is $0.0125U$. An energy shift of about $0.001U$ in this case still means either going up one or down one table row. Again, assuming that all microstates in these rows are equally probable, the likelihood that this amount of energy will be found to have shifted from the cold solid to the hot solid compared to going the "right" way is

$$\frac{\Pr(A \to B)}{\Pr(B \to A)} \approx \frac{\Omega_{AB}(U_A \approx 0.0115U)}{\Omega_{AB}(U_A \approx 0.0135U)} = \frac{2.1 \times 10^{5423}}{4.3 \times 10^{5453}} = 0.5 \times 10^{-30} \quad \text{(T5.3)}$$

We see that the probability that energy will flow from the hot object to the cold object is completely overwhelming. If we had been trying this experiment 1 billion times per second since the beginning of the universe 4×10^{17} s ago, the chance that we would have seen this much energy flow the wrong way even *once* in a system this tiny would still be only about 1 in 5000.

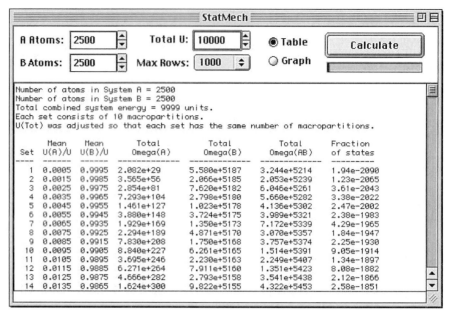

Figure T5.3
The macropartition table for a system 10 times larger than the one in figure T5.2.
Note that the multiplicity *exponents* increase by about a factor of 10!

Note that even the probability of the system's remaining somewhere in the original set of macropartitions is negligible compared to the probability of its going on to the next set closer to equilibrium:

$$\frac{\Pr(\text{no change})}{\Pr(B \to A)} \approx \frac{\Omega_{AB}(U_A \approx 0.0125U)}{\Omega_{AB}(U_A \approx 0.0135U)} = \frac{3.5 \times 10^{5438}}{4.3 \times 10^{5453}} = 0.8 \times 10^{-15} \quad \text{(T5.4)}$$

Exercise T5X.3

Using StatMech, scale the system up by a factor of 10 again. (Note that a cubic solid containing 25,000 atoms is still only about 30 atoms on a side!) Show that the probability that the system will stay in the original set of macropartitions compared to going to the next set closer to equilibrium (the ratio analogous to the one shown in equation T5.4) is only 0.5×10^{-150}. What is the value of the ratio analogous to the one in equation T5.3?

When the numbers become as large as those shown in exercise T5X.3, the idea that the system could *in principle* stay in the same macropartition (or even move to a macropartition farther away from equilibrium) has no *practical* meaning: we will never, never, *never*, even if we watch for Avogadro's number of lifetimes of the universe, observe a system larger than 10,000 atoms do anything but inexorably march toward macropartitions where the energy is more evenly distributed between its subsystems.

The computer program Equilib (also available for free download from the *Six Ideas* website) illustrates this process in action. This program models two Einstein solids in thermal contact, each of which consists of 400 fundamental oscillators (about 133 atoms' worth) arranged in a 20×20 grid of squares. Initially, the oscillators in system A have zero energy, while each oscillator in solid B has 5 units of energy. Each time step, the program scans through all 800 oscillators. If a given oscillator has any energy, the program either transfers 1 unit of energy to a randomly selected one of the oscillator's eight neighbors

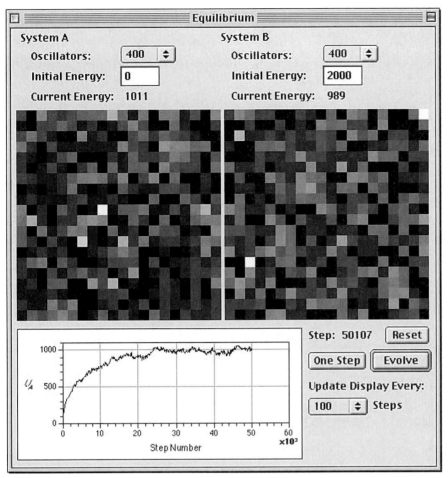

Figure T5.4
A screen shot of the Equilib program in action. Each small square represents an individual quantum oscillator that exchanges energy with its neighbors. Each square's brightness reflects the amount of energy in the corresponding oscillator.

or allows the oscillator to keep its energy (with equal probability for each of the nine possibilities). Note that solids A and B can interchange energy because oscillators along the right edge of solid A are neighbors to the oscillators along the left edge of solid B, and vice versa. After doing this for each oscillator, the program updates a display that shows each oscillator as a shaded square whose brightness indicates how many energy units the oscillator stores (see figure T5.4). The program also updates a graph showing the total energy in solid A as a function of the step number.

I strongly recommend that you play with this program; it is fascinating to watch. The point is that this program simulates a completely random interchange of energy between simple identical oscillators. It knows nothing about microstates or macrostates. Even so, you can see that (even for this very small system) U_A marches pretty much inexorably upward as time passes until the system reaches equilibrium at about step 25,000; thereafter, U_A fluctuates randomly about that most probable macropartition (where $U_A = 1000\varepsilon$). In other words, this simple system exhibits directly how random microscopic processes can lead to an essentially irreversible transfer of energy from a hot solid to a cold solid!

So, let us take a step back and consider what this means at the macroscopic level. I take a solid with a lot of thermal energy and place it in contact

The link to paradigmatic thermal process

with an otherwise identical solid with little thermal energy. What we see is that once contact is established, the random shuffling of energy between the objects leads with *virtual certainty* to a net energy flow from the hot solid to the cold solid until they reach the most probable (equilibrium) macropartition. Once the solids reach the most probable macropartition, they will stay there, again with virtual certainty, except for tiny random fluctuations (that are essentially immeasurable for large systems).

This, then, (at least in the context of Einstein solids) is the answer to the fundamental question of this unit. Why are some macroscopic processes irreversible? Specifically, why does heat always flow from hot to cold and never the other way? The answer is that *there are more microstates in that direction.* It is not because energy cannot *in principle* flow the other way. It is not because the time-reversible laws of microscopic physics for some reason do not apply here. Complex objects simply evolve by random process to the macrostates that contain the most microstates. It is really (surprisingly, wonderfully) simple!

T5.3   Irreversibility in General

The general requirements for irreversible behavior

Look back over the argument that I presented in section T5.2. The details of exactly how many microstates Einstein solids have in this macropartition as opposed to that macropartition are not really very important. Only three things have to be true for this explanation of irreversibility to work:

1. There must be a countable number of macroscopically indistinguishable microstates for every macrostate of the combined system of two subsystems.
2. Each accessible microstate of the combined system should be roughly equally probable (the fundamental assumption of statistical mechanics).
3. Each subsystem's multiplicity must (like that of a large Einstein solid) be a *very* rapidly increasing function of its thermal energy U.

The first of these requirements is a simple consequence of quantum mechanics, which requires the energy states of any bound system to be quantized. So long as we can use quantum mechanics to describe the system (which we can for all systems as far as we know), the first statement will be true. The second is the fundamental *assumption* of our model, but predictions based on that assumption do seem to be correct in a very wide range of circumstances. So let us focus on answering the following questions about the third requirement: (1) why does irreversibility necessarily follow if it is true, and (2) *is* it true for most macroscopic objects?

Why the third statement implies irreversibility

In any situation where the last statement is true, the combined system's multiplicity will exhibit an extremely sharp peak at the most probable macropartition. To see how this works, consider the graphs shown in figure T5.5. Imagine that we set the scale of the vertical axis of a graph so that the top of the scale corresponds to the value of subsystem A's multiplicity Ω_A at some arbitrarily chosen value U_0 of its thermal energy. If Ω_A increases rapidly with thermal energy, then at values of U_A greater than U_0 it will be off the scale, and at values even a bit less than U_0 it will be so much smaller as to be essentially zero. Therefore, a graph of Ω_A versus U_A will look something like that shown in figure T5.5a. Moreover, if subsystem B's multiplicity is a very rapidly increasing function of *its* internal energy U_B, then when system B is in contact with system A, its multiplicity will be a very rapidly *decreasing* function of U_A (since if U_A increases, U_B decreases). Therefore, a graph of Ω_B versus U_A will look something like that shown in figure T5.5b. The combined system's

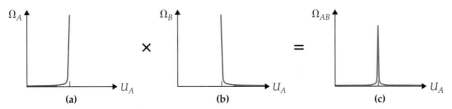

Figure T5.5
The product of rapidly increasing and rapidly decreasing functions is a spike. (In all three graphs, I have defined the vertical axis so that the value of the function at the mark on the horizontal axis is full scale on the graph's vertical axis.)

multiplicity, which is the product $\Omega_{AB} = \Omega_A\Omega_B$ of these functions, will indeed be a sharply peaked function of U_A, as shown in figure T5.5c.

So, as long as both $\Omega_A(U_A)$ and $\Omega_B(U_B)$ are *very* rapidly increasing functions, Ω_{AB} will increase so rapidly as one goes toward the peak and the numbers involved will be so huge that the probability that random processes happen to move the system *away* from the peak will be zero for all practical purposes. Moreover, the peak will be so narrow that once the combined system arrives at the vicinity of the peak macropartition, it will stay within a very narrow range with virtual certainty. Therefore, such a system will exhibit all the hallmarks of irreversible behavior.

We have seen that this statement is true for the Einstein solid, but is it true for other thermodynamic systems? It is indeed! I in fact chose the Einstein solid as our working example not only because it is relatively easy to count its microstates but also precisely because *its behavior in this regard is completely typical of the vast majority of complex systems.*

To predict *precisely* how rapidly the multiplicity of a substance increases with energy, we need to construct a detailed model of the substance, and doing this can be pretty difficult. But *qualitatively*, almost any macroscopic system behaves as an Einstein solid. Like an Einstein solid, any macroscopic system will contain a huge number of atoms or molecules, each of which can be in a variety of different energy states. The details of how a molecule's energy levels are spaced may be different, but (as long as there is no upper limit on the molecule's energy) such details do not affect the qualitative result that there will be a *huge* number of ways to distribute the system's energy among the molecules and that increasing the energy even a small amount opens up a vast number of new possibilities for distributing that energy. My point is that the argument presented in this chapter that explains the irreversible behavior of Einstein solids also *qualitatively* applies to essentially all other macroscopic systems as well.

The third statement is true for general systems

T5.4   The Definition of Entropy

To know which macropartitions of a combined system are the most probable, we need to know how many microstates the system has in that macropartition. However, as we have seen, the number of microstates available to even a tiny system can be so large as to be extremely awkward. Indeed, the number of microstates available to Einstein solids involving Avogadro's number of atoms are in the ballpark of $10^{10^{23}}$ (that is, 1 followed by Avogadro's number of zeros). Such numbers are truly unmanageable. For realistically sized objects, we need a less awkward way to talk about multiplicity.

Toward this end, we define an object's **entropy** S in a given macrostate to be

The definition of entropy

$$S \text{ (of macrostate)} \equiv k_B \ln \Omega \text{ (of macrostate)} \qquad \text{(T5.5)}$$

Purpose: This equation defines the entropy S of a thermodynamic system in a given macrostate.
Symbols: Ω is the macrostate's multiplicity; k_B is Boltzmann's constant.
Limitations: There are none; this is a definition. However, it is difficult to determine Ω exactly in many situations.

In the specific case of an Einstein solid, whose macrostate is specified by U and N, its entropy is a function of U and N:

Entropy of an Einstein solid

$$S(U, N) \equiv k_B \ln \Omega(U, N) = k_B \ln \frac{(3N + U/\varepsilon - 1)!}{(3N - 1)!(U/\varepsilon)!} \qquad \text{(T5.6)}$$

Some nice features of this definition

Why this definition? *Entropy* is really just another way of talking about the *multiplicity:* when the multiplicity is large, the entropy is large. Defining the entropy in terms of the *natural logarithm* of the multiplicity, however, serves two very useful purposes. First, it makes the awkwardly large multiplicity values easier to manage. For example, the logarithm of a multiplicity $\Omega = 10^{10^{23}}$ is about 2.3×10^{23}. While still a large number, this is much more manageable than the multiplicity itself.

Second, consider a specific macropartition of a system of two systems in thermal contact. The multiplicity of the combined system in this macropartition is $\Omega_{AB} = \Omega_A \Omega_B$, where Ω_A and Ω_B are the multiplicities of the individual systems in their specified macrostates. This implies that the total entropy of the combined system in that macropartition is conveniently the simple *sum* of the solids' individual entropies in that macrostate.

$$S_{AB} = S_A + S_B \qquad \text{(T5.7)}$$

Purpose: This equation describes the total entropy S_{AB} for a given macropartition of a system consisting of two subsystems A and B in thermal contact.
Symbols: S_A and S_B are the subsystems' entropies in that macropartition.
Limitations: There are none. This directly follows from the definitions of entropy and multiplicity.

Exercise T5X.4

Show that equation T5.7 follows from the definition of entropy and the mathematical fact that $\ln xy = \ln x + \ln y$.

We could have defined $S = \ln \Omega$ without multiplying the logarithm by k_B (and a handful textbook authors do this). This shares all the advantages of the conventional definition and in a certain sense more fundamentally captures the essence of entropy. The constant k_B just rescales $\ln \Omega$ in a way that we will see is helpful when we explore the definition of temperature in chapter T6. For now, simply note that since $k_B = 1.38 \times 10^{-23}$ J/K, a multiplicity on the order of magnitude of $10^{10^{23}}$, whose logarithm is about 2.3×10^{23}, will have a corresponding entropy $S = k_B \ln \Omega$ of about 4 J/K, a number of convenient size.

Example T5.1

Problem   Imagine a situation in which we have two Einstein solids in thermal contact with $N_A = N_B = 250$ and $U = 1000\varepsilon$. (a) Find the entropies of both individual solids and the combined system in the macropartition where $U_A = 12\varepsilon$. (b) Compare the entropy of the total system in this macropartition with that of the total system in the central macropartitions (where $\Omega_{AB} = 9.86 \times 10^{726}$). Express all entropies as a multiple of k_B.

Translation   According to figure T5.2, $\Omega_A = 7.22 \times 10^{25}$, $\Omega_B = 1.03 \times 10^{514}$, and $\Omega_{AB} = 7.42 \times 10^{539}$.

Model   The first of the entropies is easy to calculate directly: $S_A \equiv k_B \ln \Omega_A = 59.5k_B$. However, my calculator cannot handle exponents greater than 99, so I cannot calculate either $\ln \Omega_B$ or $\ln \Omega_{AB}$ directly. However, one of the properties of the natural logarithm is that $\ln x^a = a \ln x$, so $\ln 10^{514} = 514 \ln 10 = 514(2.30258)$. Moreover, $\ln xy = \ln x + \ln y$.

Solution   Therefore,

$$\ln \Omega_B = \ln 1.03 + \ln 10^{514} = 0.030 + 514(2.30258) = 1183.56 \quad \text{(T5.8)}$$

so $S_B = 1183.56k_B$. Similarly, $S_{AB} = k_B[\ln 7.42 + 539(2.30258)] = 1243.10k_B$. The entropy of the central macropartitions is $S_{AB} = k_B[\ln 9.86 + 726(2.30258)] = 1673.97k_B$.

Evaluation   This is substantially larger, as expected.

Example T5.2

Problem   Imagine that one macropartition of a combined system of two Einstein solids has an entropy of $S_1 = 432.5k_B$, while another (where the energy is more evenly divided) has an entropy of $S_2 = 546.3k_B$. How many times more likely is it that you will find the system in the second macropartition than in the first?

Model   The probability of a given macropartition being observed is proportional to its multiplicity Ω. Now, the natural log and exponential functions are inverses, so $e^{\ln x} = x$. Since the definition of entropy is $S = k_B \ln \Omega$, this means that $\Omega = e^{S/k_B}$.

Solution   Therefore, the ratio of the probabilities in this case is

$$\frac{\Omega_2}{\Omega_1} = \frac{e^{S_2/k_B}}{e^{S_1/k_B}} = \frac{e^{546.3}}{e^{432.5}} = e^{546.3-432.5} = e^{113.8} = 2.6 \times 10^{49} \quad \text{(T5.9)}$$

Therefore, we are 2.6×10^{49} times more likely to find it in the second macropartition than in the first macropartition.

T5.5   The Second Law of Thermodynamics

We have seen that a system of two objects will evolve from macropartitions with lower multiplicities (and thus lower entropy) to macropartitions with higher multiplicities (higher entropy). Moreover, once the system reaches the macropartition having the highest multiplicity (highest entropy), it will stay there. For very tiny systems, these are merely statements of probability, but for anything approaching a normal-size system, the extremely large multiplicities turn probabilities in principle to certainties in practice. Therefore, as a large system inevitably evolves from macropartitions with low multiplicities to those with larger multiplicities, *its entropy inevitably increases*. The entropy of a (sufficiently large) system will *never* be observed to decrease, because the idea that the system would evolve to a macropartition where its entropy is measurably smaller is extraordinarily improbable (see problem T5S.8 for a numerical example). The following simple law, known as the **second law of thermodynamics,** expresses the essence of this idea:

> *The entropy of an isolated system never decreases.*

This law is one of the most important and useful laws of physics and is (as the British writer C. P. Snow once asserted) one of those things a well-educated person ought to know. It expresses in a precise and quantitative way the fact of irreversibility and has a number of important implications and applications that we will study in future chapters. But we should never lose sight of the fact (as Boltzmann first argued) that this great law is a *consequence* of the more fundamental ideas presented in section T5.3.

T5.6   Entropy and Disorder

Many popular treatments link *entropy* to the concept of *disorder*. Here we have defined entropy in terms of *multiplicity*. How are these ideas related?

Think about it this way. Why does your dorm room get disorderly unless you specifically clean it up? *Because that's where the microstates are.* There are many, many more ways for your room to be disorderly than orderly. Because of this, random things that happen in your room are *far* more likely to contribute to disorder than to greater order. (What is the likelihood that an earthquake will pick up your clothes and hang them back up in your closet?)

Let us consider some more physical examples of disorder. A flowing liquid of a certain kind of atom has greater disorder than a neat crystalline solid made of the same atoms. A substance with a sufficient amount of thermal energy will therefore be a liquid instead of a solid because *that's where the microstates are:* there are many, many more microstates available to molecules if they are free to roam around than if they are confined to a solid.

Figure T5.6 shows a container with two compartments. Before mixing, each compartment contains different gases. A valve between the two compartments is then opened, allowing the gases to mix with each other. You may know that under such circumstances, the gases will *spontaneously* intermingle, just as a drop of cream put in a cup of coffee will naturally diffuse throughout the cup (although a little stirring speeds up the process). Why do the gases spontaneously mix, becoming more "disorderly" in a certain sense? *Because that's where the microstates are.* There are many, many, many more

The second law of thermodynamics

Examples of disorder linked to multiplicity

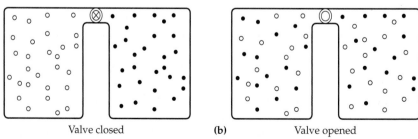

(a)   Valve closed   **(b)**   Valve opened

Figure T5.6
Spontaneous mixing of gases.

Figure T5.7
The contents of which glass have the higher entropy?

microstates available to the system when the gas molecules are free to roam over the entire container than there are when the gases are separated and confined to their side of the valve.

So in many standard situations, increased multiplicity is indeed linked with disorder. It is important to remember, however, that entropy is defined in terms of *multiplicity*, not disorder. Increasing entropy does not *necessarily* imply increasing disorder (at least *visible* disorder).

For example, consider the glass of crushed ice and the glass of water shown in figure T5.7. Which of the two glasses has a higher entropy? The crushed ice may *look* more disorderly; but since we have to add energy to the ice to convert it to water, and since the multiplicity of virtually all macroscopic systems increases dramatically when we add energy, the multiplicity (and thus the entropy) of the nice, orderly glass of water is much larger.

Another potentially confusing example is life itself. Life has been described by some as being anti-entropic, because living things take unorganized inorganic matter and transform it to the most intricate and complex order that one can imagine. Is this not inconsistent with the law of increasing entropy?

Not at all! No creature is an isolated system. In fact, every creature releases energy into its environment as it organizes matter. This energy causes the entropy of the creature's surroundings to increase far more than the creature decreases its own entropy by putting a few things in order. Indeed, thoughts in a brain, movements of a muscle, growth, development, even evolution itself should be considered examples of *entropy in action*. Sources of energy are tapped by living things and ultimately dissipated into the environment, and this natural, spontaneous flow of energy from being concentrated to dissipated is cleverly tapped by organisms to help accomplish a little organization along the way. The beautiful order of life is thus a *manifestation* of the dissipation of energy (and consequent increase in entropy of the universe) in much the same way as the orderly ticking of a clock manifests the winding down of its spring. Stars and galaxies form, the sun shines, tectonic plates move, volcanoes erupt, storms form, and flowers bloom in the spring because all systems in the universe are evolving toward the macropartitions that maximize their entropy (see figure T5.8). Every physical process in our daily life, whether associated with growth or decay, increasing order or increasing disorder, is not merely *consistent* with the second law of thermodynamics: it is a *consequence* of that law.

There are some cases, then, in which the idea of *entropy as disorder* can be misleading. Whenever there appears to you to be a contradiction between the concepts of entropy as disorder and entropy as multiplicity, you should remember that *multiplicity* is the more basic concept.

In some cases, the link with disorder is misleading

The basic meaning of entropy is multiplicity, *not* disorder

Figure T5.8
Entropy in action (the flowers, too).

TWO-MINUTE PROBLEMS

T5T.1 A hot object is placed in contact with a cold object. Heat is observed to flow spontaneously from the hot object to the cold object, but not in the other direction. According to the argument in this chapter, this is so because
A. This increases the entropies of both objects.
B. This decreases the entropy of the combined system.
C. The system will tend to evolve toward macropartitions that have more microstates.
D. *A* and *C*
E. *B* and *C*

T5T.2 Consider a system whose multiplicity is always 1, no matter how much energy you put into it. If you put a very large amount of energy into such an object (object *A*), and place it into thermal contact with an Einstein solid (object *B*) having the same number of atoms but much less energy, what will happen?
A. Energy flows from *A* to *B* until *A* has virtually no energy.
B. Energy flows from *B* to *A* until *B* has virtually no energy.
C. Energy flows from *A* to *B* until they have the same energy.
D. No energy will flow between *A* and *B* at all.
E. Something else happens (describe).

T5T.3 Consider a system comprised of two or more objects in contact. The system's total entropy in a given macropartition is always equal to the *sum* of the entropies of its parts (T or F).

T5T.4 $10^{62}/10^{60} = 100$. What is $\ln 10^{62}/\ln 10^{60}$? (Select the closest response.)
A. 100
B. 10
C. 4.6
D. 1.0

T5T.5 The entropy of a certain macropartition of a combined system is $102k_B$. The entropy of another macropartition is $204k_B$. How much more likely is the system to be in the second macropartition than the first? (Select the closest response.)
A. 2 times
B. $e^2 = 7.4$ times
C. $10^2 = 100$ times
D. $e^{102} = 2 \times 10^{44}$ times
E. $e^{102k_B} \approx e^0$; the two macropartitions have basically the same probability

T5T.6 Which system has the greater entropy, a puddle of water sitting on a table next to a chaotic scattering of salt crystals or the same amount of salt dissolved in an evenly distributed way in the puddle? Be prepared to describe your reasoning.

A. The puddle and the scattered crystals.

B. The puddle with the dissolved salt.
C. Both have the same entropy.
D. There is no way to tell without having detailed models of both the water and the salt.

HOMEWORK PROBLEMS

Basic Skills

T5B.1 Consider the system consisting of a pair of Einstein solids in thermal contact. A certain macropartition has a multiplicity of 3.7×10^{1024}, while the total number of microstates available to the system in all macropartitions is 5.9×10^{1042}. If we look at the system at a given instant of time, what is the probability that we will find it to be in our certain macropartition?

T5B.2 Consider the system consisting of a pair of Einstein solids in thermal contact. A certain macropartition has a multiplicity of 1.2×10^{346}, while the total number of microstates available to the system in all macropartitions is 5.9×10^{362}. If we look at the system at a given instant of time, what is the probability that we will find it to be in our certain macropartition?

T5B.3 Consider the system consisting of a pair of Einstein solids in thermal contact. Imagine that it is initially in a macropartition that has a multiplicity of 8.8×10^{123}. The adjacent macropartition closer to the equilibrium macropartition has a multiplicity of 4.2×10^{134}. If we look at the system a short time later, how many times more likely is it to have moved to the second macropartition than to have stayed with the first?

T5B.4 Consider the system consisting of a pair of Einstein solids in thermal contact. Imagine that it is initially in a macropartition that has a multiplicity of 7.6×10^{3235}. The adjacent macropartition closer to the equilibrium macropartition has a multiplicity of 4.1×10^{3278}. If we look at the system a short time later, how many times more likely is it to have moved to the second macropartition than to have stayed with the first?

T5B.5 An Einstein solid in a certain macrostate has a multiplicity of 3.8×10^{280}. What is its entropy (expressed as a multiple of k_B)?

T5B.6 A pair of Einstein solids in a certain macropartition has multiplicities of 4.2×10^{320} and 8.6×10^{132}. What are the entropies of each solid? What is the total entropy of the system in this macropartition? (Express entropies as multiples of k_B.)

T5B.7 Is it *really* true that the entropy of an isolated system consisting of two Einstein solids *never* decreases? (Consider a pair of very small solids.) Why is this statement more accurate for large systems than for small systems? Explain in your own words.

Synthetic

T5S.1 You ask your roommate to clean up a mess he or she made in your room. Your roommate refuses, because cleaning up the mess would violate the second law of thermodynamics, and campus security's record of your roommate's legal violations is already excessive. Gently but firmly explain why complying will not put your roommate at risk of such an infraction.

T5S.2 The classic statement of Murphy's law reads, "If something can go wrong, it will." Explain how this is really a consequence of the second law of thermodynamics. (*Hint:* What is the entropy of "wrong" in a given context compared to the entropy of "right"?)

The cover of a recent issue of *Scientific American* shows Murphy's law in action.

T5S.3 (a) Run the StatMech program for two Einstein solids in contact with $N_A = N_B = 100$ and $U = 200\varepsilon$ and answer the following questions. (1) How many times more likely is the system to be found in the center macropartition than in the extreme macropartition where $U_A = 0$ and $U_B = 200\varepsilon$? (2) What is the range of values that U_A is likely to have more than 99.98% of the time? (3) If U_A were initially to have the extreme value 0, how many times more likely is it to move to the next macropartition nearer the center than to remain in the extreme one?

(b) Answer the same questions as in (a) for a run where you scale everything up by a factor of 10, so that $N_A = N_B = 1000$ and $U = 2000\varepsilon$.

(c) Answer the same questions as in (a) for a run where $N_A = N_B = 1000$ and $U = 200\varepsilon$. Comment on the effect that increasing just the *size* of the system by a factor of 10 has on these answers.

(d) Answer the same questions as in (a) for a run where $N_A = N_B = 100$ and $U = 2000\varepsilon$. Comment on the effect that increasing just the *energy* available to the system by a factor of 10 has on these answers.

T5S.4 Consider two Einstein solids in thermal contact. The solids have different values of N but are identical in all other respects. It is *plausible*, since every atom in the combined system is identical, that in equilibrium the energy will be distributed among the solids in such a way that the average energy per atom is the same. Use StatMech to test this hypothesis in the situation where $U = 1000\varepsilon$ and N_A and N_B have various different values such that $N_A + N_B = 1000$. (Set Max Rows to 1000 so that you can see every macropartition.)

(a) Is it true in *most* cases that in the most probable macropartition the solids have energies such that the average energy per atom in each is the same? Is it strictly true in *every* case? Answer these questions by discussing the values of N_A and N_B you tested, and whether the actual most probable macropartition is the same as that predicted by the hypothesis.

(b) In any case where the hypothesis does *not* work, does increasing both N_A and N_B by a factor of 10 or 100 (but leaving U alone) yield a result more or less consistent with the hypothesis?

(c) Speculate as to the value of this hypothesis in the large-N limit.

T5S.5 (a) What is the entropy of an Einstein solid with 5 atoms and an energy of 15ε? Express your answer as a multiple of k_B.

(b) What is the entropy of an Einstein solid with 50 atoms and an energy of 100ε? Express your answer as a multiple of k_B. (*Hint:* You will find that using StatMech is by far the fastest way to calculate the multiplicity here.)

T5S.6 A certain macropartition of two Einstein solids has an entropy of $305.2k_B$. The next macropartition closer to the most probable one has an entropy of $335.5k_B$. If the system is initially in the first macropartition and we check it again later, how many times more likely is it to have moved to the other than to have stayed in the first?

T5S.7 My calculator cannot display e^x for $x > 230$. One can calculate e^x for larger values of x as follows. Define y such that $x \equiv y \ln 10$. This means that $e^x = e^{y \ln 10} = (e^{\ln 10})^y = 10^y = 10^{x/\ln 10}$. Note that we can calculate 10 raised to a non-integer power (for example, $10^{3.46}$) as follows: $10^{3.46} = 10^{3+0.46} = 10^3(10^{0.46}) = 2.9 \times 10^3$.

Use these techniques to solve the following problem. The entropy of the most probable macropartition for a certain system of Einstein solids is $6025.3k_B$, while the entropy of an extreme macropartition is only $5755.4k_B$. What is the probability of finding the system at a given time in the extreme macropartition compared to that of finding it in the most probable macropartition?

T5S.8 *In principle,* the entropy of an isolated system decreases a little bit whenever random processes cause its macropartition to fluctuate away from the most probable macropartition. We can certainly see this with small systems (see figure T5.4). But is this really a possibility for a typical macroscopic system? Imagine that we can measure the entropy of a system of two solids to within 2 parts in 1 billion. This means that we could just barely distinguish a system that has an entropy of 4.99999999 J/K (eight 9s!) from one that has 5.00000000 J/K. (This is a reasonable entropy for a macroscopic system.)

(a) Imagine that the entropy of the equilibrium macropartition is 5.00000000 J/K. Show that the approximate probability that at any given time later we will find the system in a macropartition with entropy 4.99999999 J/K (i.e., with an entropy that is only barely measurably smaller) is about $10^{315,000,000,000,000}$ times smaller than the probability we will still find it to have entropy 5.00000000 J/K. (*Hint:* See problem T5S.7.)

(b) Defend the statement that the entropy of an isolated system in thermal equilibrium *never* decreases.

Rich-Context

T5R.1 According to a reputable checkout-counter news source, space aliens give top scientists an object made of a substance that can store thermal energy but whose multiplicity actually *decreases* as its energy increases. What will happen to such an object if it is placed in thermal contact with a normal object (such as an Einstein solid)? Can this object ever be in thermal equilibrium with a normal object? If you put an object of this substance in a flame, will the flame warm it? How might you increase the thermal energy of such an object? Do you think that we can assign a meaningful temperature to such an object? Answer these questions *qualitatively,* but carefully, supporting your answers with arguments based on the ideas in this chapter. (*Hint:* What would a macropartition table look like for such an object in contact with an Einstein solid?)

Comment: This is not just a science fiction scenario: certain real physical systems can exhibit such behavior. For example, if one sets up the system described in problem T4R.1 with enough initial energy in a very low-temperature environment, it can behave in this way. In general, this kind of weird behavior is possible in systems where there is an upper limit on the energy that a molecule can hold, which is not the case in Einstein solids and most other substances.

T5R.2 (Adapted from Kittel and Kroemer, *Thermal Physics*, 2d ed., p. 53.) The writer Aldous Huxley is reported to have said that "six monkeys, set to strum unintelligently on typewriters for millions of years would be bound in time to write all of the books in the British Museum." This statement is in fact completely *false*: Huxley has been misled by an incorrect intuition about the character of extremely large numbers.

Let us set ourselves a much less challenging task. Imagine that we have 10 billion monkeys (somewhat more than the human population of the earth) typing diligently at the rate of 2 characters per second since the universe began 5×10^{17} s ago. Instead of requiring an entire library, we will settle for a single typed version of *Hamlet*. Let us guess that *Hamlet* has approximately 10^5 characters.

(a) Assume that the typewriters used by the monkeys have 26 letters and 10 punctuation characters (space, carriage return, period, comma, colon, semicolon, quotation mark, apostrophe, dash, question mark) for a total of 36 characters. We will ignore the distinction between capital letters and small letters. The probability that any given character is the first character in *Hamlet* is thus 1/36. The probability that this character and the next are the first *two* characters of *Hamlet* is $(1/36)(1/36) = (1/36)^2$. Argue, then, that the probability that any given sequence of 10^5 random characters is *Hamlet* is $10^{-155,630}$. (*Hint:* $\log x^a = a \log x$, where log is the base-10 logarithm.)

(b) Argue that the probability that *Hamlet* would be randomly typed by any of our army of monkeys since the beginning of the universe is about $10^{-155,602}$. (*Hint:* Remember that *any* key typed by *any* monkey could in principle be the first character in the play.) This probability is *zero* in any practical sense: Even our huge bevy of monkeys will never, never, *never* be able to type *Hamlet* at random.

Advanced

T5A.1 We can estimate the approximate size of a system's fluctuations around the equilibrium macropartition as follows. When the number of molecules N is extremely large, many substances have multiplicities of the form $\Omega \approx CU^{aN}$, where U is the substance's thermal energy, C is a quantity that depends on the type of substance and N but not on U, and a is a substance-dependent constant of the order of magnitude of 1. Imagine two identical objects in thermal contact.

a. Argue that we can write the multiplicity of a combined system with total energy U as

$$\Omega_{AB} \approx B(1 - x^2)^{aN} \qquad \text{where } x \equiv \frac{U_A - \frac{1}{2}U}{\frac{1}{2}U} \quad \text{(T5.10)}$$

and B is another quantity that does not depend on U. Note that the variable x here specifies how far we are from equilibrium. If $x = 0$, then $U_A = \frac{1}{2}U$, which is the equilibrium macropartition, and $x = \pm 1$ corresponds to the extreme macropartitions $U_A = U$ and $U_A = 0$, respectively. Note that Ω_{AB} has a peak at $x = 0$ and falls off very rapidly as the absolute value of x grows.

b. Show the value of x where Ω_{AB} has fallen to one-half of its peak value is

$$x_{1/2} = \pm \frac{1}{\sqrt{2aN}} \qquad \text{(T5.11)}$$

(*Hint:* Argue that one can use the binomial approximation to calculate the binomial.)

c. Assuming that $a \approx 1$ and $N \approx 10^{24}$, estimate very roughly how far a typical fluctuation will carry the energy of either system away from its equilibrium value $\frac{1}{2}U$, expressed as a fraction of that equilibrium value.

ANSWERS TO EXERCISES

T5X.1 The probability of the 1:199 macropartition is 3.3×10^{-41}.

T5X.2 The number of seconds since the beginning of the universe is very roughly $(1.4 \times 10^{10} \text{ y})(3 \times 10^7 \text{ s/y}) = 4 \times 10^{17}$ s. If we peek 1 billion times per second, we will have peeked about $(4 \times 10^{17})(10^9) = 4 \times 10^{26}$ times since the beginning of the universe. Since the probability that any peek yields $U_A = 0$ is roughly 4×10^{-43}, the probability that we will have seen this value of U_A in any one of our peeks is roughly $(4 \times 10^{-43})(4 \times 10^{26}) \approx 2 \times 10^{-16}$. This is a small enough probability that I would be willing to bet my life on this not happening.

T5X.3 The probability of U_A moving $0.001U$ from $U_A = 0.0125U$ toward smaller energy as opposed to larger energy in this case is

$$\frac{\Pr(A \to B)}{\Pr(B \to A)} \approx \frac{\Omega_{AB}(U_A \approx 0.0115U)}{\Omega_{AB}(U_A \approx 0.0135U)}$$

$$= \frac{4.1 \times 10^{54,282}}{2.3 \times 10^{54,587}} = 1.8 \times 10^{-305} \quad \text{(T5.12)}$$

since $54,282 - 54,587 = -305$.

T5X.4 The multiplicity of any given macropartition of the combined system is

$$\Omega_{AB} = \Omega_A \Omega_B \qquad \text{(T5.13)}$$

If we take the natural log of both sides, we get

$$\ln \Omega_{AB} = \ln \Omega_A \Omega_B = \ln \Omega_A + \ln \Omega_B \qquad \text{(T5.14)}$$

Multiplying both sides of this equation by k_B and using the definition of S given by equation T5.5, we get equation T5.7.

Temperature and Entropy

Chapter Overview

Introduction

In this chapter, we will use the concepts presented in chapter T5 to develop a new and more fundamental definition of temperature that does not depend on the arbitrary choice of a practical thermoscope. We will spend much of the rest of the unit exploring the practical implications of this definition.

Section T6.1: The Definition of Temperature

The total entropy $S_{TOT} = S_A + S_B$ of a system consisting of two objects A and B is a maximum (by definition!) when the system is in its equilibrium (most probable) macropartition. This means that in equilibrium

$$0 = \frac{dS_{TOT}}{dU_A} = \frac{dS_A}{dU_A} + \frac{dS_B}{dU_A} = \frac{dS_A}{dU_A} - \frac{dS_B}{dU_B} \quad \Rightarrow \quad \frac{dS_A}{dU_A} = \frac{dS_B}{dU_B} \quad \text{(T6.3)}$$

(The next-to-last step follows because any energy that object A gains comes at the expense of object B, implying that $dU_A = -dU_B$.) By the zeroth law, the two objects' temperatures (by definition!) must be equal in that macropartition as well. The mathematical consequence of these statements (as discussed in the section) is that the derivative dS/dU of an object's entropy S with respect to its thermal energy U must be equal to some function $f(T)$ of that object's temperature T. The function $1/T$ is the simplest choice for $f(T)$ consistent with the idea that heat spontaneously flows from an object with a large value of T to an object with a lower value of T. Therefore we *define* an object's temperature as follows:

$$\frac{1}{T} \equiv \frac{dS}{dU} \text{ (when } V, N, \text{ etc. are constant)} \equiv \frac{\partial S}{\partial U} \quad \text{(T6.6)}$$

Purpose: This equation defines an object's absolute temperature T in terms of the rate at which the object's entropy S increases with its internal energy U.

Limitations: The derivative must be evaluated while holding all of the object's other macroscopic variables fixed (this is what the partial derivative symbol $\partial S/\partial U$ means).

Note that since in principle we can *calculate* an object's multiplicity and thus its entropy if we have a decent model, we can calculate an object's temperature from fundamental principles without referring to a standard thermoscope.

Section T6.2: Consistency with the Old Definition

Is this new definition of temperature at least approximately consistent with the old gas thermoscope definition? We saw in chapter T2 that at sufficiently high temperatures, the thermal energy U of a monatomic solid satisfies the relationship $U \approx 3Nk_BT$, where T in this equation was the old gas thermoscope temperature. A calculation based on equation T6.6 yields the same result, strongly suggesting that the new definition of temperature is at least approximately the same as the old definition.

Section T6.3: A Financial Analogy

Objects in thermal contact behave as people involved in financial transactions do if we say that

Energy	$\longleftrightarrow$	Money
Entropy	$\longleftrightarrow$	Happiness
Temperature	$\longleftrightarrow$	Generosity

and assume that financial transactions between people always serve to increase the happiness of everyone involved. Just as the general happiness increases if money flows from a more generous person to a less generous person, the entropy of a thermodynamic system increases if energy flows from a high-temperature object to a lower-temperature one.

Most normal objects have entropies and temperatures that increase with energy, just as normal people's happiness and generosity increase when they receive money. It is possible to find physical systems that are "miserly" (whose temperature decreases as their energy increases) or "enlightened" (whose entropy increases as their energy decreases). The financial analogy helps us think clearly about how such exotic objects might behave when interacting with normal objects.

Section T6.4: The Boltzmann Factor

We define a (thermal) **reservoir** to be an object so large that it can provide or absorb any energy likely to be exchanged in a situation of interest without undergoing a significant change in its temperature. Consider a small quantum system in thermal contact with such a reservoir. The probability that the quantum system will be in a quantum state with energy E is proportional to the multiplicity of the combined system. But the multiplicity of the quantum system in a given quantum state is 1 by definition, so the multiplicity of the combined system is the same as the *reservoir's* multiplicity. The reservoir's multiplicity will decrease as E increases because the quantum system's energy comes at the expense of the reservoir's energy. A short calculation involving the definition of temperature implies that the reservoir's multiplicity (and thus the state's probability) in fact decreases *exponentially* with E:

$$\Pr(E) = \frac{1}{Z}e^{-E/k_B T} = \frac{e^{-E/k_B T}}{\sum_{\text{all states}} e^{-E_i/k_B T}} \tag{T6.23}$$

Purpose: This equation describes the probability that a small system in thermal contact with a reservoir at absolute temperature T will be in a quantum state (i.e., a microstate) with energy E.

Symbols: E_i is the energy of the ith quantum state of the small system, Z is a constant of proportionality, and k_B is Boltzmann's constant.

Limitations: The reservoir must be large enough that it can provide the small system with any energy it is likely to have without suffering a significant change in its temperature T.

Note: We call $e^{-E/k_B T}$ the **Boltzmann factor**.

Section T6.5: Some Simple Applications

Equation T6.23 has a large number of useful applications. This section illustrates how we can use the equation to calculate the probabilities of the two different configurations of a certain molecule in a solution at room temperature, and the probability that a hydrogen atom's electron will be in its first excited state at room temperature.

T6.1   The Definition of Temperature

I stated in chapter T1 that statistical mechanics makes it possible to define absolute temperature in a very fundamental and general way *without* referring to any particular thermoscope. My goal in this section is to present this new and powerful definition of temperature.

Consider again our famous *paradigmatic thermal process*, as illustrated in figure T6.1. A hot object is brought into contact with a cold object. Subsequently heat flows from the hot object to the cold object, decreasing the energy (and thus entropy) of the hot object and increasing the energy (and thus entropy) of the cold object.

Assuming that the two objects have a large number of molecules, the combined system will evolve inexorably toward the most probable macropartition and subsequently remain there, in *thermal equilibrium*. The thermal equilibrium macropartition will therefore be the macropartition with the greatest number of microstates, and therefore the greatest total entropy $S_{TOT} = S_A + S_B$.

Now, the combined system's entropy S_{TOT} is a function of the system's macropartition, which in a pure heat transfer process (where the volume V, the number of molecules N, and other macroscopic characteristics of the interacting objects are held constant) is completely determined by the objects' energies U_A and U_B. Actually, we only need to know U_A to determine the macropartition, since $U_B = U - U_A$, where U is the fixed total energy of the combined system. The macropartition where S_{TOT} is maximum is specified by the value of U_A such that

$$0 = \frac{dS_{TOT}}{dU_A}. \tag{T6.1a}$$

(This is the usual way of finding the maximum of a function.) Since the system's total entropy $S_{TOT} = S_A + S_B$, we can rewrite this as follows:

$$0 = \frac{d(S_A + S_B)}{dU_A} = \frac{dS_A}{dU_A} + \frac{dS_B}{dU_A} \tag{T6.1b}$$

by the sum rule of differential calculus. Now, according to the chain rule,

$$\frac{dS_B}{dU_A} = \frac{dS_B}{dU_B}\frac{dU_B}{dU_A} = \frac{dS_B}{dU_B}(-1) = -\frac{dS_B}{dU_B} \tag{T6.2}$$

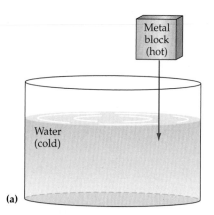

(a)

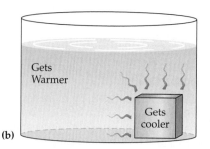

(b)

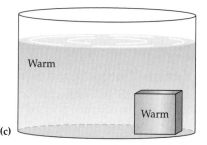

(c)

Figure T6.1
The paradigmatic thermal process.

since $U_B = U - U_A$. (If you prefer a more informal argument, note that since any thermal energy gained by A is lost by B, $dU_A = -dU_B$. Plugging this into dS_B/dU_A yields $-dS_B/dU_B$.) If we plug equation T6.2 into equation T6.1b and rearrange things a bit, we find that the two objects will be in thermal equilibrium if and only if

$$\frac{dS_A}{dU_A} = \frac{dS_B}{dU_B} \qquad (T6.3)$$

The necessary condition for equilibrium

This is illustrated in figure T6.2.

Exercise T6X.1

Verify equation T6.3.

Note that we can calculate the quantities on either side of equation T6.3 *without reference to the other object:* we simply take the derivative of an object's entropy S with respect to its *own* thermal energy U (holding its other macroscopic properties constant). This equation thus tells us that when two objects are in the equilibrium macropartition, the quantity dS/dU calculated for each object must be the same.

But the fundamental meaning of temperature (as the zeroth law of thermodynamics asserts) is that the *temperature* of two objects is the same in equilibrium. Therefore, the quantities dS/dU and temperature T must be linked in some one-to-one relationship. The most general way to describe this relationship is to say that for a given object

$$f(T) \equiv \frac{dS}{dU} \qquad (T6.4)$$

where $f(T)$ is some as yet unknown function of the object's temperature T.

Different choices for $f(T)$ simply amount to different ways of linking dS/dU (the quantity most directly connected to equilibrium) to numerical temperature values. *In principle*, we could choose $f(T)$ to be anything we like as long as $f(T)$ has a unique value for every value of T (so that two objects having the same temperature necessarily have the same value of dS/dU and vice versa): our choice would simply define a new temperature scale. *In practice*, though, physicists elected to choose $f(T)$ so that the new scale we define using equation T6.4 corresponds at least reasonably closely to the historical constant-volume gas thermoscope scale described in chapter T1.

One desirable feature of that original scale is that heat flows from a hot object (i.e., one with a large value of T) to a cold object (one with a low T). As we have seen, energy flows spontaneously between two objects A and B in thermal contact because this allows the combined system's *total* entropy S_{TOT} to increase. Imagine that object A is the cold object, so that in an infinitesimal heat transfer process it gets a positive increment in energy dU_A from the other object. Since S_{TOT} must increase in this process, dS_{TOT}/dU_A must also be positive. Therefore, we have

$$0 < \frac{dS_{\text{TOT}}}{dU_A} = \frac{dS_A}{dU_A} + \frac{dS_B}{dU_A} = \frac{dS_A}{dU_A} - \frac{dS_B}{dU_B} \equiv f(T_A) - f(T_B) \quad (T6.5)$$

Exercise T6X.2

Explain the reasoning behind each step in the sequence of equalities in equation T6.5.

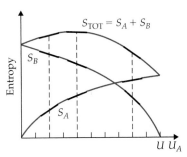

Figure T6.2
The system's total entropy is maximum where the slopes of the entropy curves for parts A and B become equal in magnitude and opposite in sign (at about $U_A = 0.4U$ in the case shown). When this is so, the entropy gained by A when it gets a bit of energy dU_A from B is exactly balanced by the entropy lost by B. If the slopes are not equal, the system can gain entropy by transferring energy from one part to the other.

So for heat to flow from B to A, we must have $f(T_A) > f(T_B)$. But since object A here is the *cold* object, we see that as T increases, the value of $f(T)$ should get *smaller*. The *simplest* definition of $f(T)$ consistent with this requirement is $f(T) = 1/T$. So (following Boltzmann) we define temperature T as follows:

The fundamental definition of temperature

$$\frac{1}{T} \equiv \frac{dS}{dU} \text{ (when } V, N, \text{ etc. are constant)} \equiv \frac{\partial S}{\partial U} \qquad \text{(T6.6)}$$

Purpose: This equation defines an object's absolute temperature T in terms of the rate at which the object's entropy S increases with its internal energy U.

Limitations: The derivative must be evaluated while holding all of the object's other macroscopic variables fixed (this is what the partial derivative symbol $\partial S/\partial U$ means).

(You may have encountered **partial derivatives** already in unit E. If you have not, do not be stressed out by this new notation. A partial derivative is really the same thing as an ordinary derivative, except that the funny notation reminds us to hold everything that S might depend on except for U constant when we evaluate the derivative with respect to U.)

Since in principle we can count the number of microstates available to an object in a given macropartition, and from that determine how $S \equiv k_B \ln \Omega$ varies with internal energy, we can calculate any object's temperature T from very fundamental physical quantities. *This is an extremely important equation!* The rest of this unit will be almost exclusively focused on exploring its implications.

Exercise T6X.3

Show that having Boltzmann's constant k_B in the definition $S \equiv k_B \ln \Omega$ ensures that the temperature T defined by equation T6.6 has units of kelvins.

T6.2   Consistency with the Old Definition

Equation T6.6 defines temperature T in a manner consistent with the convention that heat flows from objects with large values of T to objects with low values of T. However, this in itself does not ensure that the temperature scale so defined is consistent with the old gas thermoscope scale. In this section, we will see evidence suggesting that these two scales are in fact equivalent.

We saw in chapter T2 that at sufficiently high temperatures, the thermal energy U of a monatomic solid satisfies the relationship

$$U = 3Nk_B T \qquad \Rightarrow \qquad \frac{1}{T} = \frac{3Nk_B}{U} \qquad \text{(T6.7)}$$

where T in this equation was the old gas thermoscope temperature. But we know the multiplicity of an Einstein solid, so in principle we can use equation T6.6 to *calculate* $1/T$ for a monatomic solid (modeled as an Einstein solid) as predicted by the new definition. If we get the same result, it will be strong evidence that our new definition of temperature coincides with the old.

Starting with equation T6.6, applying the definition $S = k_B \ln \Omega$, the chain rule, and the fact that the derivative of $\ln x$ is $1/x$, we get

$$\frac{1}{T} = \frac{\partial S}{\partial U} = k_B \frac{\partial}{\partial U}(\ln \Omega) = k_B \frac{1}{\Omega} \frac{\partial \Omega}{\partial U} = k_B \frac{1}{\Omega} \frac{\partial \Omega}{\varepsilon \, \partial q} \tag{T6.8}$$

where $q \equiv U/\varepsilon$ is the total number of fundamental energy units that the solid contains. According to equation T4.7, the multiplicity of an Einstein solid is

$$\Omega = \frac{(q + 3N - 1)!}{q!(3N - 1)!} \tag{T6.9}$$

We can estimate the value of $\partial \Omega / \partial q$ by computing $\Delta \Omega / \Delta q$ for the smallest possible value of Δq (which is $\Delta q = 1$) while holding N fixed:

$$\frac{\partial \Omega}{\partial q} \approx \frac{\Omega(q + 1) - \Omega(q)}{1} = \frac{[(q + 1) + 3N - 1]!}{(q + 1)!(3N - 1)!} - \frac{(q + 3N - 1)!}{q!(3N - 1)!} \tag{T6.10}$$

But $(n + 1)! \equiv (n + 1) \cdot n \cdot (n - 1) \cdots 1 = (n + 1) \cdot n!$. Plugging this into equation T6.10 and using equation T6.9, we find that

$$\frac{\partial \Omega}{\partial q} = \frac{q + 3N}{q + 1}\Omega - \Omega = \Omega \left[\frac{3N - 1}{q + 1}\right] \approx \Omega \left[\frac{3N}{q}\right] \tag{T6.11}$$

assuming both $3N$ and q are large compared to 1.[†]

Exercise T6X.4

Fill in the missing steps in equation T6.11.

If we plug this back into equation T6.8 and use $q \equiv U/\varepsilon$, we find that according to our new definition of temperature,

$$\frac{1}{T} = \frac{k_B}{\varepsilon} \frac{1}{\Omega}\left(\Omega \frac{3N}{q}\right) = k_B \frac{3N}{\varepsilon q} = \frac{3Nk_B}{U} \tag{T6.12}$$

which is the result we found empirically for the gas thermoscope temperature. This strongly suggests that the new definition of temperature coincides with the old. (We will see even more compelling evidence in chapter T7.)

Example T6.1

Problem In the limit that $q \gg 3N \gg 1$, one can show by an entirely different argument (see problem T6A.1) that the multiplicity of an Einstein solid is given approximately by

$$\Omega(U, N) \approx \left(\frac{eU}{3N\varepsilon}\right)^{3N} \tag{T6.13}$$

where $e = 2.718$. Show that this approximation also implies that $U \approx 3Nk_B T$.

[†]Technically, the difference in equation T6.10 will only be a good approximation to a derivative if its value does not change significantly as q changes by $\Delta q = 1$. Note that according to equation T6.11, Ω changes by a factor of $(q + 3N)/(q + 1)$ as q increases by 1, and since the derivative is proportional to Ω, it will increase by this factor also. So for the change in the difference to be small when $\Delta q = 1$, we must also have $q \gg 3N$, so that the ratio $(q + 3N)/(q + 1) \approx 1$. This means that this calculation will only work when the solid is hot enough that each oscillator, on average, has many units of energy.

Solution Equation T6.13 implies that

$$\ln \Omega = \ln \left(\frac{eU}{N\varepsilon} \right)^{3N} = 3N \ln \frac{eU}{N\varepsilon} = 3N(\ln e + \ln U - \ln N - \ln \varepsilon) \quad \text{(T6.14)}$$

The definition of temperature then implies that

$$\frac{1}{T} = \frac{\partial S}{\partial U} = \frac{\partial}{\partial U}(k_B \ln \Omega) \approx 3Nk_B \frac{\partial}{\partial U}(\ln e + \ln U - \ln N - \ln \varepsilon)$$

$$= 3Nk_B \left(0 + \frac{1}{U} + 0 + 0 \right) = \frac{3Nk_B}{U} \quad \text{(T6.15)}$$

Multiplying both sides of this equation by UT yields $U \approx 3Nk_B T$.

T6.3   A Financial Analogy

My friend Daniel Schroeder, in his text *Thermal Physics* (San Francisco: Addison Wesley Longman, 2000), uses the following light-hearted analogy to make the definition of temperature clearer. An isolated system of objects that constantly exchange energy so as to maximize the system's total entropy is analogous to a community of people who exchange money so as to maximize not their own happiness but the community's total happiness. A *cold* object has a large value of dS/dU according to our definition of temperature, implying that its entropy increases rapidly as it receives energy and drops dramatically when it gives up energy. We would describe the analogous person whose happiness increases rapidly when receiving money and decreases rapidly when giving money as being "greedy." Conversely, a hot object has a small value of dS/dU, implying that its entropy does not change much when it gains or loses energy. We would call the analogous person who is not greatly excited by receiving money and not greatly perturbed to give away money "generous." If you put a generous person in contact with a greedy person, the generous person will end up paying some money to the greedy person to increase the community's happiness, just as a hot object will give up energy to a cold object to increase the system's entropy. Note also that a community's happiness can be increased by a judicious redistribution of money (even though the total amount of money does not increase) in the same way that a system's entropy can be increased by a redistribution of conserved energy among its parts. So we see that an analogy in which

The core of the analogy

Energy	$\longleftrightarrow$	Money
Entropy	$\longleftrightarrow$	Happiness
Temperature	$\longleftrightarrow$	Generosity

can help us intuitively understand the link between temperature, entropy, and energy.

Types of people compared to types of systems

A *normal* person becomes more generous as he or she receives money, and analogously the temperature of a *normal* object becomes larger as it receives energy. If we draw a graph of the entropy S of a normal object versus its energy U, we get something like that shown in figure T6.3a: note that the slope dS/dU decreases (implying that the temperature T increases) as U increases.

However, there is nothing in physics that prevents an object's temperature from decreasing when you add energy. Indeed, systems such as stars or star clusters that are bound by gravitational interactions behave in this way.

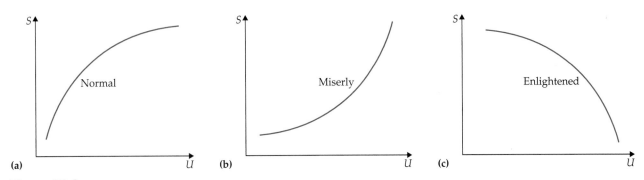

Figure T6.3

Entropy versus energy graphs for a normal object, a miserly object, and an enlightened object.

Adding energy to a star causes it to expand, and this ends up putting more energy into the gravitational potential energy of expansion than was added originally, requiring the star's temperature to decrease. We would call the analogous person who gets more greedy as you give him or her money miserly. Figure T6.3b shows a graph of S versus U for a miserly physical system. Fortunately, genuinely miserly people are rare in society, and miserly physical systems are also pretty rare.

Rarer still are those enlightened individuals whose happiness actually increases when they give money away. Figure T6.3c shows graph of S versus U for the analogous physical system. Systems whose entropy decreases with increasing energy would have *negative* temperature, according to our definition! Yet just as an enlightened person would gladly give away all his or her money and a normal person would be happy to receive it, such an "enlightened" physical system will give up all its energy to a normal system, and so in a certain sense is "hotter" than any normal object with positive temperature! This kind of a situation is extremely counterintuitive to say the least (because we are so used to normal objects), but physical systems like this *can* be constructed with some effort. (The system discussed in problem T4R.1 is enlightened in this sense when $U > \frac{1}{2}N\varepsilon$, meaning that more than one-half of its atoms are in the single excited state available to each atom.)

We will deal essentially entirely with "normal" physical objects in this course. Even so, it is good for your flexibility of mind to know that more exotic systems do exist. The financial analogy, even though it is a bit silly, helps one think clearly about *all* kinds of systems. We will also return to the financial analogy in chapter T9.

T6.4   The Boltzmann Factor

Much of the rest of this text will be devoted to exploring consequences of the fundamental equation $1/T = \partial S/\partial U$. In the remaining sections of this chapter as well as chapter T7, we will consider its application to a special situation that is nonetheless very useful in a broad range of circumstances. Consider a small system in thermal contact with a very much larger system (see figure T6.4 for a schematic representation). For example, the small system could be a single gas molecule in contact with the rest of the gas in a room, or a single solute molecule in a solution. The question we will answer in this section is, When a small system is in contact with a large system with a given temperature T, how does the probability that the small system (whatever it is) will be in a given quantum state (or microstate) depend on that state's energy E?

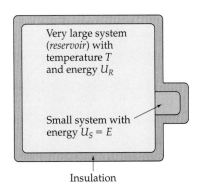

Figure T6.4

In the remainder of this chapter we will consider a very small system in thermal contact with a large system (both isolated from everything else).

Definition of a **reservoir**

We can answer this question fairly easily in the limit that the large system is very much larger than the small system. We have seen before that the change dU in an object's energy is related to its change in temperature dT by the equation $dU = mc\,dT$, where m is the object's mass and c is its specific heat. We can see from this that if a system is sufficiently massive, it can supply or absorb a significant amount of thermal energy and yet suffer only a negligible change in temperature. A **reservoir** is a system so large that it can absorb or supply any energy of interest in a given situation without suffering a measurable change in temperature. In what follows, we will assume that the large system is large enough to be a reservoir with respect to the typical energies that the small system might absorb or emit.

Exercise T6X.5

Imagine that an atom in 1 mol of helium gas absorbs 100 electronvolts (abbreviated as eV) of energy from the rest of the gas (this is a huge amount of energy for a single helium atom). How much does the temperature of the rest of the gas decrease? Can the rest of the gas be considered a reservoir for that helium atom?

Derivation of the Boltzmann factor

Now consider two different quantum states of the small system, one with energy E_0 and one with energy $E_1 > E_0$. What is the probability that the small system will be in the higher energy state relative to that for the lower energy state? According to the fundamental assumption of statistical mechanics, the ratio of the probabilities is equal to the ratio of the combined system's multiplicities in the two cases:

$$\frac{\Pr(E_1)}{\Pr(E_0)} = \frac{\Omega_{SR,1}}{\Omega_{SR,0}} \tag{T6.16}$$

where $\Omega_{SR,0}$ and $\Omega_{SR,1}$ are the combined system's multiplicities when the small system has energies E_0 and E_1, respectively. In general, the combined-system multiplicity is the product of the multiplicities of both systems: $\Omega_{SR} = \Omega_S \Omega_R$. However, in both cases, we are interested in the probability of the small system being in a *single* quantum state, so $\Omega_S = 1$ by definition. So

$$\frac{\Pr(E_1)}{\Pr(E_0)} = \frac{1 \cdot \Omega_{R1}}{1 \cdot \Omega_{R0}} = \frac{\Omega_{R1}}{\Omega_{R0}} \tag{T6.17}$$

where Ω_{R1} is the *reservoir's* multiplicity when the *small* system has energy E_1 and so on. Now, the reservoir's entropy, by definition, is $S_R = k_B \ln \Omega_R$, so $\Omega_R = e^{S_R/k_B}$. Plugging this into equation T6.17, we get

$$\frac{\Pr(E_1)}{\Pr(E_0)} = \frac{e^{S_{R1}/k_B}}{e^{S_{R0}/k_B}} = e^{(S_{R1}-S_{R0})/k_B} = e^{\Delta S_R/k_B} \tag{T6.18}$$

where ΔS_R is the change in the *reservoir's* entropy when the small system's energy increases from E_0 to E_1. We are assuming that the combined system is isolated from the rest of the world, though, so the total energy of the combined system must be conserved. This means that if the small system's energy increases by $\Delta U_S = E_1 - E_0$, the energy of the reservoir *decreases* by the same amount:

$$\Delta U_R = -(E_1 - E_0) \tag{T6.19}$$

Here is where the reservoir assumption enters

Finally, note that the fundamental equation $1/T = \partial S/\partial U$ tells us that if the small system's volume does not significantly change when its energy increases and the reservoir is large enough that changing its energy by ΔU_R does not change its temperature T measurably, then $\Delta S_R \approx \Delta U_R / T$. Plugging these results into equation T6.18, we get

$$\frac{\Pr(E_1)}{\Pr(E_0)} = e^{\Delta U_R / k_B T} = e^{-(E_1 - E_0)/k_B T} = \frac{e^{-E_1/k_B T}}{e^{-E_0/k_B T}} \tag{T6.20}$$

Since this relationship has to be true for all pairs of small-system quantum states, it follows that the probability that a small system in contact with a reservoir at temperature T will be in a quantum state with any energy E is given by

$$\Pr(E) \propto e^{-E/k_B T} \quad \text{or} \quad \Pr(E) = \frac{1}{Z} e^{-E/k_B T} \tag{T6.21}$$

where $1/Z$ is a constant of proportionality that must (by definition) be the same for all quantum states of the small system.

Since the *total* probability of the small system being in *some* quantum state must be 1, it follows that

$$1 = \sum_{\text{all states}} \Pr(E_i) = \sum_{\text{all states}} \frac{1}{Z} e^{-E_i/k_B T} = \frac{1}{Z} \sum_{\text{all states}} e^{-E_i/k_B T}$$

$$\Rightarrow \quad Z = \sum_{\text{all states}} e^{-E_i/k_B T} \tag{T6.22}$$

where E_i is the energy of the ith small-system quantum state. Therefore, the absolute probability of any small-system quantum state having energy E is

$$\Pr(E) = \frac{1}{Z} e^{-E/k_B T} = \frac{e^{-E/k_B T}}{\displaystyle\sum_{\text{all states}} e^{-E_i/k_B T}} \tag{T6.23}$$

The probability that the small system is in quantum state with energy E

Purpose: This equation describes the probability that a small system in thermal contact with a reservoir at absolute temperature T will be in a quantum state (i.e., a microstate) with energy E.

Symbols: E_i is the energy of the ith small-system quantum state, Z is a constant of proportionality, and k_B is Boltzmann's constant.

Limitations: The reservoir must be large enough that it can provide the small system with any energy it is likely to have without suffering a significant change in its temperature T.

Notes: We call $e^{-E/k_B T}$ the **Boltzmann factor.**

This important equation has *many* interesting applications, only some of which we can explore in this chapter and chapter T7. Note that it follows directly from the definitions of entropy and temperature, the fundamental assumption that all microstates of the combined system are equally likely, and the assumption that the reservoir's temperature is essentially unchanged as the small system's energy fluctuates.

Equation T6.23 implies that the probability of a small system being in a quantum state with energy E decreases exponentially as E increases. This makes sense if you remember that increasing the energy of the small system means decreasing the energy in the reservoir, which decreases the number of microstates available to the reservoir. Since the probability of the small system's having energy E in this situation is entirely determined by the number of microstates available to the *reservoir* (see equation T6.17), it makes sense

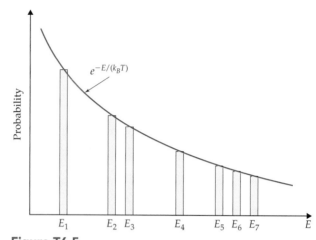

Figure T6.5

No matter what the energies of the small system's quantum states are, their probability decreases exponentially with increasing energy.

that the probability should decrease as the small system's energy increases. Equation T6.23 asserts that this probability in fact decreases exponentially (see figure T6.5).

We can display this exponential dependence for systems of harmonic oscillators using the StatMech program introduced in chapter T4. First, note that the probability that the combined system is in a macropartition where the small system is in a quantum state with energy E is $\Omega_{SR}/\Omega_{\text{TOT}}$, where Ω_{SR} is the macropartition's multiplicity and Ω_{TOT} is the total number of microstates available to the combined system. Therefore equation T6.23 quite generally implies that

$$\frac{S}{k_B} \equiv \ln \Omega_{SR} = \ln\left[\Omega_{\text{TOT}}\text{Pr}(E)\right] = \ln\left(\frac{\Omega_{\text{TOT}}}{Z}e^{-E/k_B T}\right) = \text{constant} - \frac{E}{k_B T}$$

(T6.24)

where $S \equiv k_B \ln \Omega_{SR}$ is the entropy of the combined system in that macropartition. So if we plot S/k_B versus the small system's energy E, we should get a straight line with slope $-1/k_B T$.

To get StatMech to display such a graph, we need to do several things. First, we need to set up solid A so that it consists of a single *oscillator* (not a three-oscillator "atom") and the multiplicity of solid A is always 1. We can do this by selecting Count Oscillators under StatMech's Options menu and typing 1 for the number of oscillators in solid A. Solid B will be the reservoir, so type in a fairly large number here.

Second, we want to display a graph of S/k_B versus $E = U_A$, and we are most interested in the part of the graph where $U_A \ll U_B$, since if system A steals a significant fraction of system B's energy, its temperature will change and therefore we cannot consider it to be a good reservoir. If you select S for Small U(A) under the GraphTypes menu, StatMech will plot S/k_B versus U_A for only the first 20 macropartitions in the macropartition table. If we choose Max Rows to be 1000 and the system's total energy U to be 1000ε (where ε is the difference between oscillator energy levels), then the values of U_A displayed on the graph will never exceed 2% of U.

Figure T6.6 shows the resulting graph for a small system consisting of a single oscillator in thermal contact with a large "reservoir" consisting of

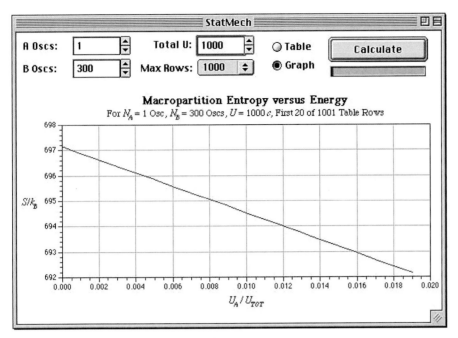

Figure T6.6
A graph of $S = k_B \ln \Omega$ versus the small system's energy $E = U_A$. The straight line indicates that Ω decreases exponentially with E.

300 oscillators having a total energy of 1000ε. We see that the graph is indeed a nearly straight line with negative slope.

Exercise T6X.6

Imagine that the difference between oscillator energy levels is $\varepsilon = 0.1$ eV. From the slope of the graph in figure T6.6, estimate the temperature of the reservoir when it has an internal energy of nearly 1000ε.

T6.5   Some Simple Applications

Equation T6.23 has a very broad range of applications that we can only begin to explore in this chapter. In this section, we will consider a few illustrative examples.

Example T6.2

Problem   Imagine that a certain molecule has two configurations, one with an energy E_0 and another with energy $E_1 = E_0 + 0.052$ eV. If we have N molecules in a solution at room temperature, roughly what fraction will be in each configuration?

Model   We will assume the solution here is large enough compared to a single molecule that it qualifies as a reservoir and that each configuration corresponds to a unique quantum state for the molecule. We will also assume that room temperature is $T = 23°C = 298$ K.

Solution   Given these assumptions, we can in principle use equation T6.23 to compute the probability of each configuration. For example,

$$\Pr(E_0) = \frac{1}{Z}e^{-E_0/k_B} \tag{T6.25a}$$

where

$$Z \equiv \sum_n e^{-E_n/k_B T} = e^{-E_0/k_B T} + e^{-E_1/k_B T} \tag{T6.25b}$$

We could calculate this fairly easily if we knew E_0, but we do not. However, we do know that $E_1 - E_0 = 0.052$ eV. If we pull out a factor of $e^{-E_0/k_B T}$ from the sum for Z, we find that

$$Z = e^{-E_0/k_B T}\left(1 + \frac{e^{-E_1/k_B T}}{e^{-E_0/k_B T}}\right) = e^{-E_0/k_B T}\left(1 + e^{-(E_1 - E_0)/k_B T}\right) \tag{T6.26}$$

Plugging this into equation T6.25, we find that

$$\Pr(E_0) = \frac{e^{-E_0/k_B T}}{e^{-E_0/k_B T}\left(1 + e^{-(E_1 - E_0)/k_B T}\right)} = \frac{1}{1 + e^{-(E_1 - E_0)/k_B T}} \tag{T6.27}$$

In our particular case,

$$k_B T = (1.38 \times 10^{-23} \text{ J/K})(298 \text{ K})\left(\frac{1 \text{ eV}}{1.60 \times 10^{-19} \text{ J}}\right) = 0.026 \text{ eV} \tag{T6.28}$$

which in turn implies that

$$\frac{E_1 - E_0}{k_B T} = \frac{0.052 \text{ eV}}{0.026 \text{ eV}} = 2.0 \tag{T6.29a}$$

$$\Pr(E_0) = \frac{1}{1 + e^{-(E_1 - E_0)/k_B T}} = \frac{1}{1 + e^{-2.0}} = 0.88 \tag{T6.29b}$$

We can calculate the probability of the other configuration in the same way:

$$\Pr(E_1) = \frac{e^{-E_1/k_B T}}{Z} = \frac{e^{-E_1/k_B T}}{e^{-E_0/k_B T}\left(1 + e^{-(E_1 - E_0)/k_B T}\right)}$$

$$= \frac{e^{-(E_1 - E_0)/k_B T}}{1 + e^{-(E_1 - E_0)/k_B T}} = \frac{e^{-2.0}}{1 + e^{-2.0}} = 0.12 \tag{T6.30}$$

Evaluation   Note that these two probabilities add up to 1, as they must (since only the two configurations are possible). [Indeed, we could have used this to calculate $\Pr(E_1)$ without going through the math in the last equation.]

Example T6.2 illustrates an important feature of problems involving Boltzmann factors: *Only the differences between energy levels in the small system have any meaning.* One can show quite generally (see problem T6S.6) that if we add an overall constant to all the energies of the small-system quantum states, we do not change any of the probabilities.

Example T6.3

Problem   Consider a hydrogen atom on the sun's surface, whose temperature is about 5800 K. What is the probability that such an atom will be in the $n = 2$ energy level, expressed as a fraction of the probability that it is in the ground state ($n = 1$) energy level?

Translation and Model   We will assume that the sun is large enough to be considered a reservoir compared to a hydrogen atom (an excellent approximation in this case!). We cannot use $Pr(E_2) = e^{-E_2/k_BT}/Z$ to calculate the probability here, since in order to calculate $Z \equiv \sum e^{-E_n/k_BT}$ we would have to sum over all states in the hydrogen atom, and there are an infinite number of such states. However, the problem only asks for the probability of the $n = 2$ level *relative* to the $n = 1$ level, so we can use equation T6.20:

$$\frac{Pr(E_2)}{Pr(E_1)} = \frac{e^{-E_2/k_BT}}{e^{-E_1/k_BT}} = e^{-(E_2-E_1)/k_BT} \qquad (T6.31)$$

According to unit Q, the allowed energies E_n for the hydrogen atom are

$$E_n = \frac{E_1}{n^2} \quad \text{where } E_1 = -13.6\,\text{eV} \quad \text{and} \quad n = 1, 2, 3, \ldots \quad (T6.32)$$

Since we know n for both levels, we can compute the difference $E_2 - E_1$; and since we also know that $T = 5800$ K, we have enough information to solve the problem.

Solution   The energy difference in question is

$$E_2 - E_1 = \frac{E_1}{2^2} - E_1 = (-13.6\,\text{eV})\left(\frac{1}{4} - 1\right) = +10.2\,\text{eV} \qquad (T6.33)$$

The value of k_BT here is

$$k_BT = (1.38 \times 10^{-23}\,\text{J/K})(5800\,\text{K})\left(\frac{1\,\text{eV}}{1.60 \times 10^{-19}\,\text{J}}\right) = 0.50\,\text{eV} \quad (T6.34)$$

Therefore, the ratio of the probabilities in this case is

$$\frac{Pr(E_2)}{Pr(E_1)} = e^{-10.2\,\text{eV}/0.50\,\text{eV}} = e^{-20.4} = 1.38 \times 10^{-9} \qquad (T6.35)$$

Actually, this is the probability of a hydrogen atom's being in any *single* $n = 2$ quantum state. There are actually four distinct quantum states of the hydrogen atom having the same energy E_2 (having different orbital angular momentum components) but only one having energy E_1. So the probability that the atom will be in *any* of the four $n = 2$ levels is actually $4(1.38 \times 10^{-9}) = 5.5 \times 10^{-9}$ times that of its being in the ground state.

Evaluation   Therefore, for every billion hydrogen atoms in the ground state on the sun's surface, about 5.5 atoms will be in the first excited state. This is relevant information for astronomers, because only hydrogen atoms already in the $n = 2$ level can absorb *visible* photons that also have enough energy to kick the atoms into the $n = 3$, $n = 4$, $n = 5$, or $n = 6$ levels. Since these are the *only* transitions that produce hydrogen absorption lines in the visible part of a star's spectrum, the fraction of hydrogen atoms that are ready to make such transitions determines how prominent these absorption lines are.

Exercise T6X.7

If we include electron spin, there are actually two distinct quantum states in the $n = 1$ energy level and eight in the $n = 2$ level. Argue that including this information does *not* change the probability ratio just calculated.

TWO-MINUTE PROBLEMS

T6T.1 As the energy in a normal object goes to zero, the slope of a graph of its entropy S versus thermal energy U
 A. Will approach infinity
 B. Will approach some positive value less than infinity
 C. Will approach zero
 D. Will approach some negative value less than minus infinity
 E. Will approach negative infinity
 F. Depends entirely on the object's properties

T6T.2 Certain physical objects have an entropy that increases from zero when its thermal energy is $U = 0$ to some maximum value when $U = \frac{1}{2}U_{max}$, and then goes back to zero when $U = U_{max}$ (as shown in the drawing below). The greatest thermal energy U that such an object could have when in thermal equilibrium with a normal object is closest to
 A. 0
 B. $\frac{1}{4}U_{max}$
 C. $\frac{1}{2}U_{max}$
 D. $\frac{3}{4}U_{max}$
 E. U_{max}
 F. Other (specify)

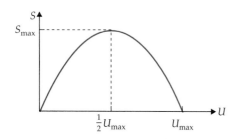

T6T.3 If an object has a multiplicity Ω that *decreases* as its thermal energy U increases, its temperature would
 A. Be always zero
 B. Be always fixed
 C. Increase with U
 D. Decrease with U
 E. Be negative
 F. Do otherwise (specify)

T6T.4 Imagine that an atom has two energy levels separated by some energy difference ΔE. When such atoms are placed in thermal contact with a reservoir at temperature T, there is roughly 1 atom in the higher state for every 4 in the lower state. What is the approximate ratio of ΔE to $k_B T$?
 A. $\frac{1}{4}$
 B. 4
 C. ln 4
 D. −ln 4
 E. $e^{-1/4}$
 F. e^{-4}
 T. Other (specify)

T6T.5 Imagine that an atom has exactly two energy levels separated by some energy difference ΔE. When such atoms are in contact with a reservoir at a temperature T_1, the ratio of the number of atoms in the higher level to the number in the lower level is $\frac{1}{4}$. If we increase the temperature to T_2, the numerical value of this ratio
 A. Increases
 B. Decreases
 C. Remains the same

T6T.6 The quantity Z in the equation $\Pr(E) = Z^{-1}e^{-E/k_B T}$ depends on the temperature of the reservoir (T or F).

HOMEWORK PROBLEMS

Basic Skills

T6B.1 An object's entropy is measured to increase by 0.1 J/K when we add 35 J of energy. What is its approximate temperature? (Assume that the object's temperature does not change much when we add the 35 J.)

T6B.2 A certain Einstein solid's entropy changes from $305.2k_B$ to $338.1k_B$ when we add 1 unit ε of energy. What is the value (and units) of $k_B T/\varepsilon$ for this solid? If $\varepsilon = 1.0$ eV, what is its temperature T?

T6B.3 Does it make sense to talk about the temperature of a vacuum? If so, how could you measure or calculate it? If not, why not?

T6B.4 It is sometimes said that $k_B T$ at room temperature is about $\frac{1}{40}$ eV. For what temperature T would this be exactly correct? By about what percent is this off at $T = 298$ K?

T6B.5 The star Vega in the constellation Lyra has a surface temperature of about 9500 K. Estimate the probability that a hydrogen atom on Vega's surface is in an $n = 2$ energy level as a fraction of the probability of its being in the ground $n = 1$ level.

T6B.6 Imagine that an atom has exactly two energy levels whose energy difference is 0.015 eV. At what temperature will there be two atoms in the higher state for every three in the lower state?

T6B.7 Imagine that an atom has exactly three energy states, with energies 0.020 eV, 0.040 eV, and 0.060 eV, respectively. Out of every 1000 atoms about how many are in each state?

Synthetic

T6S.1 In section T6.1, I argued on fairly fundamental grounds that $dS/dU = f(T)$. In principle, we could define $f(T)$ to be anything that we like: this would amount to *defining* temperature and its scale. Still, *some* definitions would violate deeply embedded preconceptions about the nature of temperature. For example, the *simplest* definition of temperature would be $dS/dU = T_{new}$. Show that this definition
(a) would imply that T_{new} has units of K^{-1} and
(b) would imply that heat would flow spontaneously from objects with low T_{new} to objects with high T_{new}. This would imply that objects with low values of T_{new} are *hot*, while objects with high values T_{new} are *cold* (we might want to call T_{new} so defined *coolness* instead of *temperature*). While we *could* define temperature in this way, it would really fly in the face of convention (if not intuition).
(c) If we did define coolness T_{new} in this way, what ordinary temperature T would an object with absolutely zero coolness ($T_{new} = 0$) have? What about something that is infinitely cool ($T_{new} = \infty$)?

T6S.2 Imagine that the entropy of a certain substance as a function of N and U is given by the formula $S = Nk_B \ln U$. Using the definition of temperature, show that the thermal energy of this substance is related to its temperature by the expression $U = Nk_B T$.

T6S.3 For an ideal monatomic gas, it turns out (as we will see in chapter T8), that the multiplicity of a system with N atoms, thermal energy U, and volume V is roughly $\Omega(U, V, N) = CV^N U^{3N/2}$, where C is a constant that depends on N alone. Use this information and the definition of temperature to determine how the thermal energy of a monatomic gas depends on its temperature.

T6S.4 Imagine that the multiplicity of a certain substance is given by $\Omega(U, N) = Ne^{\sqrt{NU/\varepsilon}}$, where ε is some unit of energy. How would the energy of an object made out of this substance depend on its temperature? Would this be a "normal" substance in the sense defined in section T6.3?

T6S.5 Consider an Einstein solid having $N = 20$ atoms.
(a) What is the solid's temperature when it has an energy of 10ε, assuming that $\varepsilon = \hbar\omega \approx 0.02$ eV? Calculate this directly from the definition of temperature by finding S at 10ε and 11ε, computing $dS/dU \approx [S(11\varepsilon) - S(10\varepsilon)]/\varepsilon$, and then applying the definition of temperature. (You will find that

your work will go faster if you use StatMech to tabulate the multiplicities.)
(b) How does this compare with the result from the formula $U = 3Nk_B T$ (which is only accurate if N is large and $U/3N\varepsilon \gg 1$)?
(c) If you have access to StatMech, repeat for $N = 200$ and $U = 100\varepsilon$. [*Hint:* If your calculator cannot handle numbers in excess of 10^{100}, use the fact that $\ln(a \times 10^b) = \ln a + b \ln 10$.]

T6S.6 Prove that the numerical value of the probability given by equation T6.23 is unchanged if we add a constant value E_0 to the energy of each energy state available to the small system.

T6S.7 (Adapted from Schroeder, *Thermal Physics*.) A water molecule can vibrate in many ways, but the "flexing" mode, where the angle between the OH bonds oscillates about its central value of 104°, is the vibrational mode whose excited energies are the smallest. We can model this vibrational mode as if it were a simple one-dimensional quantum harmonic oscillator (mathematically, the situations are equivalent); so as we saw in unit Q, the energy levels are $\frac{1}{2}\hbar\omega$, $(3/2)\hbar\omega$, $(5/2)\hbar\omega$, and so on, where ω is the angular frequency of the oscillation. The measured value of ω for this mode is 3.0×10^{14} s^{-1}. Calculate the absolute probabilities that a water molecule at room temperature will be in its flexing ground state and its first two excited states. (*Hint:* Calculate Z by adding up as many Boltzmann factors you think you need to get a result accurate to within four significant digits. This will not involve very much work.)

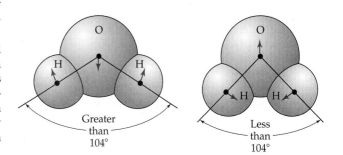

A water molecule's flexing mode of oscillation.

T6S.8 The ground ($n = 1$) state of a hydrogen atom is actually two different quantum states with slightly different energies. Because of a magnetic interaction between the proton and electron, the quantum state where the proton and electron spins are aligned has an energy a few microelectronvolts above the energy of the state where the spins are anti-aligned.
(a) If the hydrogen atom undergoes a transition between these states, it emits a photon with a characteristic wavelength of 21 cm. (Astronomers can detect such photons with a radio telescope and thus determine the distribution of hydrogen gas

in the sky.) What is the energy difference between the two states?

(b) If a hydrogen cloud in intergalactic space has a temperature of about 2.9 K (about equal to the temperature of the cosmic background radiation), about what fraction of hydrogen atoms in the $n = 1$ level will be in the aligned state (and thus be capable of emitting the 21-cm photons)?

A map of the 21-cm radio emission from the spiral galaxy M51.

T6S.9 (Adapted from Schroeder, *Thermal Physics*.) At very high temperatures (as could be found in the very early universe), we can model the proton and neutron as being two different quantum states of the same quanton, which we might call a "nucleon." A nucleon can easily be converted from one to the other of its two states by interacting with energetic electrons and/or neutrinos, which were also abundant at the time. The neutron's rest energy is higher than the proton's by about 1.3 megaelectronvolts (MeV). At the time when the universe had a temperature of about 10^{11} K, what was the approximate ratio of protons to neutrons?

Rich-Context

T6R.1 (Adapted from Schroeder, *Thermal Physics*.) Cold interstellar molecules often contain CN (cyanogen) molecules, whose three first excited rotational quantum states all have an energy of 4.7×10^{-4} eV above that of the single ground rotational state. Studies of absorption spectrum of starlight traveling through one cloud done in 1941 implied that for every 10 CN molecules in the ground state, roughly 3 were in the first excited state (i.e., an average of one in each of the three individual rotational states with the same energy).

(a) To explain these data, astronomers at the time argued that the molecules might be in thermal equilibrium with some "reservoir" at a well-defined temperature. If this is so, what is the approximate temperature T of this reservoir?

(b) In the mid-1960s, physicists discovered that the universe was filled with cosmic background radiation. Anything exposed to this radiation would behave as if it were in thermal contact with a reservoir with a temperature of 2.7 K. Could this be the reservoir that explains these data?

T6R.2 Imagine that aliens deliver into your hands two identical objects made of substances whose multiplicities increase linearly with thermal energy, something like $\Omega = aNU/\varepsilon$, where ε is some energy unit and a is some constant. Answer the following questions about these objects.

(a) Do they have a well-defined temperature? If so, how does this temperature depend on the objects' thermal energy?

(b) If these objects are placed in thermal contact, will energy spontaneously flow from hot to cold? Will the objects eventually come into equilibrium at a certain common temperature? (*Hint:* I suggest drawing a graph of Ω_{AB} versus macropartition. This will also help in the next part.)

(c) How will the size of random fluctuations in the energies of these objects compare to those for two normal objects placed in thermal contact?

(d) In what other ways (if any) will these objects behave differently from normal objects? (For example, what happens if you take one of these objects and slice it in two?)

Advanced

T6A.1 We saw in chapter T4 that the multiplicity of an N-atom Einstein solid in a macrostate where it has total energy U is

$$\Omega(N, U) = \frac{(3N + q - 1)!}{(3N - 1)! \, q!} \tag{T6.36}$$

where $q \equiv U/\varepsilon$ and ε is the energy difference between energy levels in the individual quantum oscillators. When N is large, we can ignore the 1 in comparison to $3N$, so this becomes

$$\Omega(N, U) \approx \frac{(3N + q)!}{(3N)! \, q!} \tag{T6.37}$$

Now, *Stirling's approximation* asserts that when m is large,

$$\ln(m!) \approx m \ln m - m \qquad \text{when } m \gg 1 \tag{T6.38}$$

a. Using this approximation, *assuming* that the ratio $U/(3N\varepsilon) \gg 1$ (that is, that the object's total energy is large enough that there is plenty of energy to give each oscillator many ε's worth of energy), and using the approximation $\ln(1 + x) \approx x$ when $x \ll 1$, show that $\ln \Omega$ is approximately

$$\ln \Omega \approx 3N \ln \frac{U}{3N\varepsilon} + 3N \qquad (T6.39)$$

b. Argue that if we take the exponential of both sides of this equation, we get

$$\Omega \approx \left(\frac{eU}{3N\varepsilon} \right)^{3N} \qquad (T6.40)$$

as stated in equation T6.13.

ANSWERS TO EXERCISES

T6X.1 If we substitute equation T6.2 into equation T6.1b, we get

$$0 = \frac{dS_A}{dU_A} + \frac{dS_B}{dU_A} = \frac{dS_A}{dU_A} - \frac{dS_B}{dU_B} \qquad (T6.41)$$

Adding dS_B/dU_B to both sides yields

$$\frac{dS_A}{dU_A} = \frac{dS_B}{dU_B} \qquad (T6.42)$$

T6X.2 The first equality follows from equation T6.1b, the second from substituting the result of equation T6.2, and the third from the definition of temperature given in equation T6.4.

T6X.3 Ω is unitless and k_B has units of joules per kelvin, so $S \equiv k_B \ln \Omega$ has units of joules per kelvin. Since U has units of joules, dS/dU has units of

$$\frac{J/K}{J} = \frac{1}{K} \qquad (T6.43)$$

Since $dS/dU \equiv 1/T$, T will have units of kelvins.

T6X.4 Since $(n + 1)! = (n + 1) \cdot n!$, it follows that

$$\frac{[(q + 1) + 3N - 1]!}{(q + 1)!(3N - 1)!} = \frac{(q + 1 + 3N - 1)(q + 3N - 1)!}{(q + 1)q!(3N - 1)!}$$

$$= \frac{q + 3N}{q + 1} \frac{(q + 3N - 1)!}{q!(3N - 1)!}$$

$$= \frac{q + 3N}{q + 1} \Omega \qquad (T6.44)$$

This explains the first equality. Extracting the common factor of Ω and putting things over a common denominator, we get

$$\frac{q + 3N}{q + 1} \Omega - \Omega = \Omega \left[\frac{q + 3N}{q + 1} - 1 \right]$$

$$\approx \Omega \left[\frac{q + 3N - (q + 1)}{q + 1} \right]$$

$$= \Omega \left[\frac{3N - 1}{q + 1} \right] \qquad (T6.45)$$

In the last step in equation T6.11, we are assuming that $3N$ and q are both so large that we can neglect the ones in the numerator and denominator.

T6X.5 *Translation and model:* The total energy of a monatomic ideal gas such as helium is $U = (3/2)Nk_BT$, where N is the number of atoms, T is the gas's temperature, and k_B is Boltzmann's constant. Since we know N and k_B, we can use this to calculate the change dT in the gas's temperature when its energy changes by dU.

$$dU = \frac{3}{2}Nk_B \, dT \quad \Rightarrow \quad dT = \frac{2 \, dU}{3Nk_B}$$

$$\Rightarrow \quad dT = \frac{2(100 \text{ eV})(1.6 \times 10^{-19} \text{ J/eV})}{3(6.02 \times 10^{23})(1.38 \times 10^{-23} \text{ J/K})}$$

$$\approx 1.3 \times 10^{-18} \text{ K} \qquad (T6.46)$$

We can safely treat this as being negligible; so, yes, we can treat the rest of the gas as being a reservoir.

T6X.6 The graph implies that the value S_R/k_B drops in a nearly straight line by about 5.0 as the value of U_A increases from 0 to $0.019U_{TOT} = 19\varepsilon$ (and the reservoir energy decreases by the same amount). Since $\varepsilon = 0.01$ eV, the slope $k_B^{-1}(dS_R/dU_R) = -5.0/(-0.19 \text{ eV}) = 26.3/\text{eV}$ when U_A is small. The reservoir's temperature is therefore given by

$$\frac{1}{T} = \frac{dS_R}{dU_R} = k_B \frac{1}{k_B} \frac{dS_R}{dU_R} = k_B \frac{26.3}{\text{eV}}$$

so that

$$T = \frac{1 \text{ eV}}{26.3k_B} = \frac{1.60 \times 10^{-19} \text{ J}}{26.3(1.38 \times 10^{-23} \text{ J/K})} = 441 \text{ K} \qquad (T6.47)$$

T6X.7 If we count the electron spin orientations, the ratio for the probability will be

$$\frac{8e^{-20.4}}{2} \quad \text{instead of} \quad \frac{4e^{-20.4}}{1} \qquad (T6.48)$$

These clearly represent the same ratio.

T7

Some Mysteries Resolved

Chapter Overview

Introduction

This chapter launches a three-chapter subunit exploring applications of the concept of entropy and the fundamental definition of temperature. In this particular chapter, we use the Boltzmann factor introduced in chapter T6 to answer questions about solids and ideal gases posed at the end of chapter T2.

Section T7.1: The Maxwell-Boltzmann Distribution

The probability that a gas molecule will be in a quantum state where it has speed v is proportional to its Boltzmann factor $e^{-E/k_BT} = e^{-(1/2)mv^2/k_BT}$. If we make a straightforward assumption about how molecular velocities are distributed, then it turns out that the number of gas molecule quantum states having speeds lying in a tiny range of width dv centered on speed v is proportional to $v^2\, dv$. Therefore, the probability that a molecule's speed lies in such a range is proportional to $(v^2\, dv)e^{-mv^2/2k_BT}$. We can determine the constant of proportionality by requiring that the molecule have a total probability of 1 of having a speed between zero and infinity. The result is

$$\text{Pr(speed within } dv \text{ centered on } v) = \mathscr{D}(v)\,\frac{dv}{v_P} \qquad (T7.6a)$$

$$\text{where} \quad \mathscr{D}(v) = \frac{4}{\pi^{1/2}}\left(\frac{v}{v_P}\right)^2 e^{-(v/v_P)^2} \quad \text{and} \quad v_P \equiv \left(\frac{2k_BT}{m}\right)^{1/2} \qquad (T7.6b)$$

Purpose: This equation describes the probability that a molecule in a gas with temperature T has a speed within a range of width dv centered on speed v.

Symbols: m is the molecule's mass, k_B is Boltzmann's constant, and v_P is the constant with units of speed defined in equation T7.6b.

Limitations: The value of dv must be small compared to v_P, since $\mathscr{D}(v)$ varies significantly over any range that is a significant fraction of v_P.

Notes: The function $\mathscr{D}(v)$ is (a unitless form of) the **Maxwell-Boltzmann distribution function** for molecular speeds in a gas. Note also that the function $\mathscr{D}(v)$ is maximum when $v = v_P$.

To find the probability for speed ranges that are a significant fraction of the most probable speed v_P, we must integrate equation T7.6.

Section T7.2: Counting Velocity States

This section carefully argues that if we treat the gas molecules as independent quantons in a box (using the model developed in unit Q), then we can prove that the velocities associated with the possible quantum states of each molecule are indeed distributed as assumed in section T7.1.

Section T7.3: The Average Energy of a Quantum System

We can calculate the average energy of a general quantum system as follows:

$$E_{\text{avg}} = E_0 \Pr(E_0) + E_1 \Pr(E_1) + E_2 \Pr(E_2) + \cdots = \sum_{\text{all } n} E_n \Pr(E_n) \qquad (\text{T7.17})$$

If the quantum system is in contact with a reservoir at temperature T, the nth quantum state's probability is $Z^{-1} e^{-E_n/k_B T}$, so

$$E_{\text{avg}} = \sum E_n \left(\frac{e^{-E_n/k_B T}}{Z} \right) = \frac{\sum E_n e^{-E_n/k_B T}}{\sum e^{-E_n/k_B T}} \qquad (\text{T7.19})$$

Purpose: This equation describes how we can calculate the average energy E_{avg} of a quantum system in thermal contact with a reservoir at temperature T.

Symbols: E_n is the energy of the quantum system's nth quantum state, k_B is Boltzmann's constant, and $Z \equiv \sum e^{-E_n/k_B T}$. The sums are over *all* the system's states.

Limitations: This equation only works for a system in contact with something large enough to be considered a reservoir for the energies the system might typically have.

If a macroscopic object consists of N identical quantum systems (where N is large), then we can consider any given single system to be in contact with a reservoir consisting of the others. The object's total thermal energy U is then simply $N E_{\text{avg}}$.

Section T7.4: Application to Einstein Solids

The sums in equation T7.19 can have an infinite number of terms, but as E_n becomes large, $e^{-E_n/k_B T}$ becomes *very* small, so we can ignore terms beyond a certain point. If we can express the energy of the system's nth quantum state in the form $E_n = \varepsilon f(n)$, where ε is some constant with units of energy and $f(n)$ is a known function of n, the program EBoltz can calculate E_{avg} using equation T7.19 and display it as a function of the temperature ratio T/T_ε, where $T_\varepsilon \equiv \varepsilon/k_B$ the **characteristic temperature** associated with the system's characteristic energy ε. The program can also calculate the quantum system's heat capacity dE_{avg}/dT as a function of T/T_ε.

If the quantum system in question is a simple harmonic oscillator (where $E_n = \varepsilon n$), then the graph of E_{avg} versus T/T_ε is very small when $T/T_\varepsilon \ll 1$, but approaches $k_B T$ when $T/T_\varepsilon \gg 1$. This means that at high temperatures, the total energy U of an Einstein solid consisting of $3N$ oscillators approaches $3Nk_B T$, as we have seen before. Examining a graph of the heat capacity dE_{avg}/dT also helps explain some of the details of table T2.2.

Section T7.5: Energy Storage in Gas Molecules

These graphs also apply to the average vibrational energy of a diatomic molecule. However, ε is so large for diatomic molecules that $T \ll T_\varepsilon$ at room temperature, meaning that a diatomic molecule stores very little vibrational energy at room temperature (vibrational energy storage is "switched off"). Physically, this is so because molecular collisions typically do not have nearly enough energy at room temperature to bump a diatomic molecule to its first excited vibrational state, so it remains "frozen" in its zero-energy vibrational ground state.

The graphs for rotational energy storage in diatomic molecules are qualitatively similar. The corresponding characteristic rotational energy ε for typical diatomic molecules is small enough that $T_\varepsilon = 2\,\text{K}$, so at normal temperatures $T \gg T_\varepsilon$ and rotational energy storage is fully "switched on." The exception is hydrogen, for which rotational energy storage switches on at about 150 K. These results explain the step-like behavior of hydrogen's heat capacity, shown in figure T2.2.

Most other quantum systems behave in a qualitatively similar manner.

T7.1   The Maxwell-Boltzmann Distribution

In this section, we will use the Boltzmann factor introduced in chapter T6 to answer the following question: What is the probability that a molecule in a gas at a temperature T will be moving at a certain speed v? We know from chapter T2 about the average squared speed $[v^2]_{avg}$ of molecules in a gas, but we do yet not know anything about how speeds are distributed *around* this average.

We can consider a single molecule to be a quantum system in thermal contact with a reservoir consisting of the remaining molecules in the gas. Equation T6.23 then tells us that the probability that our molecule will be in a quantum state where it has speed v will be given by

$$\text{Pr(state with speed } v) \propto e^{-E/k_B T} = e^{-(1/2)mv^2/k_B T} \tag{T7.1}$$

This gives the probability that the molecule will be in a *quantum state* with speed v, but does not yet give us the probability that it simply *has* speed v, because we don't know how many different quantum states are consistent with the molecule having that speed. It turns out that the number of quantum states available to a molecule with a speed within a range of speeds of infinitesimal width dv centered on v is proportional to $v^2\,dv$. We can see this qualitatively as follows.

Each possible molecular velocity is described by a set of three numbers $[v_x, v_y, v_z]$. On a three dimensional graph whose axes are v_x, v_y, and v_z, the set of three numbers for any particular velocity defines a point. On such a graph, the set of velocity points whose corresponding speeds $(v_x^2 + v_y^2 + v_z^2)^{1/2}$ have some specific common value v defines a spherical surface centered on the graph's origin whose radius is v and whose surface area is $4\pi v^2$. The set of velocity points whose corresponding speeds are within a range dv centered on this value therefore occupies a spherical shell of radius v, thickness dv, and thus volume $4\pi v^2\,dv$ on this graph. Now, let us *assume* (as a hypothetical model) that the points corresponding to quantum-mechanically possible molecular velocity states are evenly distributed throughout the volume of a graph like this. (I will justify this assumption in more depth in section T7.2.) If this simple model is true, the number of quantum states satisfying the constraint that the molecule's speed be within a range dv centered on some value v will be proportional to the shell's volume $4\pi v^2\,dv$ and thus to $v^2\,dv$, as claimed.

Given this model, the probability that the molecule will be in *some* quantum state (we do not mind which) consistent with its speed being within an infinitesimal range dv centered on some speed v is thus

$$\text{Pr(within } dv \text{ around } v) \propto v^2\,dv\,e^{-mv^2/2k_B T} \tag{T7.2}$$

Note that the quantity $2k_B T/m$ has units of $(J/K)(K/kg) = m^2/s^2$, so $v_P \equiv (2k_B T/m)^{1/2}$ is a constant with units of speed (I'll explain the subscript shortly). We can write our expression for the probability in terms of this constant as follows:

$$\text{Pr(within } dv \text{ around } v) \propto v^2\,dv\,e^{-(v/v_P)^2} \propto \left(\frac{v}{v_P}\right)^2 \left(\frac{dv}{v_P}\right) e^{-(v/v_P)^2} \tag{T7.3}$$

The beauty of the last version is that it is clearly a unitless number. Since the probability is also a unitless number, the constant of proportionality involved in the last expression will be unitless, too. We can determine the constant of proportionality by requiring that the total probability that the molecule has *some* speed (*any* speed) be 1 (see problem T7S.9): the result turns out to be $4/\pi^{1/2}$. The complete expression for the probability is therefore

$$\text{Pr(speed within } dv \text{ centered on } v) = \mathscr{D}(v)\frac{dv}{v_P} \qquad \text{(T7.4a)}$$

where

$$\mathscr{D}(v) = \frac{4}{\pi^{1/2}}\left(\frac{v}{v_P}\right)^2 e^{-(v/v_P)^2} \quad \text{and} \quad v_P \equiv \left(\frac{2k_BT}{m}\right)^{1/2} \qquad \text{(T7.4b)}$$

Purpose: This equation describes the probability that a molecule in a gas with temperature T has a speed within a range dv centered on speed v.

Symbols: m is the molecule's mass, k_B is Boltzmann's constant, and v_P is the constant with units of speed defined in equation T7.4b.

Limitations: dv must be small compared to v_P, since $\mathscr{D}(v)$ varies significantly over any range that is a significant fraction of v_P.

Notes: The function $\mathscr{D}(v)$ is (a unitless form of) the **Maxwell-Boltzmann distribution function** for molecular speeds in a gas. Note that v_P does not depend on v or dv, but it does depend on temperature T and the molecular mass m.

The Maxwell-Boltzmann distribution

Direct measurements of molecular velocities show that equation T7.4 does an excellent job of modeling those velocities.

Figure T7.1 shows a graph of the Maxwell-Boltzmann distribution function. Note that at small values of v/v_P, the exponential part of the function is essentially equal to 1, so the function rises pretty much as $(v/v_P)^2$. For large values of v, the exponential takes over and reduces the probability back to zero. You can pretty easily show (see problem T7S.1) that the most probable speed, *i.e.*, the speed where $\mathscr{D}(v)$ is largest, is simply v_P (hence the subscript).

Exercise T7X.1

Show $v_P = 395$ m/s for oxygen at 300 K (1 mol of oxygen has a mass of 32 g).

The function $\mathscr{D}(v)$ is not symmetric about this peak but has a long tail trailing off to $v = \infty$, so the average speed v_{avg} is a somewhat higher than v_P. It is

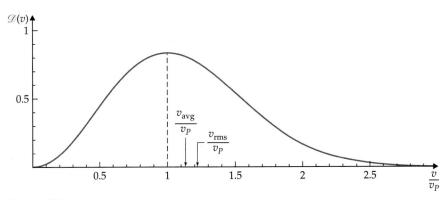

Figure T7.1
A graph of the Maxwell-Boltzmann distribution function $\mathscr{D}(v)$ as a function of v/v_P.

tricky to evaluate (see problem T7A.1), but it turns out to be

$$v_{\text{avg}} = \sqrt{\frac{8k_B T}{\pi m}} = \frac{2}{\sqrt{\pi}} \sqrt{\frac{2k_B T}{m}} = 1.13 v_P \qquad (T7.5)$$

For oxygen at 300 K, $v_{\text{avg}} = 1.13(395 \text{ m/s}) = 446 \text{ m/s}$.

The root-mean-square speed $v_{\text{rms}} \equiv ([v^2]_{\text{avg}})^{1/2}$ more heavily weights large speeds than v_{avg} does, so it is higher still. We already know from $\frac{1}{2}m[v^2]_{\text{avg}} = (3/2)k_B T$ that

$$v_{\text{rms}} = \sqrt{\frac{3k_B T}{m}} = \sqrt{\frac{3}{2}} \sqrt{\frac{2k_B T}{m}} = 1.22 v_P \qquad (T7.6)$$

Exercise T7X.2

Show that this follows from $\frac{1}{2}m[v^2]_{\text{avg}} = (3/2)k_B T$.

For oxygen at 300 K, $v_{\text{rms}} = 1.22(395 \text{ m/s}) = 482 \text{ m/s}$.

These three quantities represent different ways of expressing the characteristic speed of molecules in a gas. While they are not identical, v_P and v_{rms} differ from the average speed by only a bit more than 10%.

Example T7.1

Problem About how many molecules in 1 mol of nitrogen gas at $T = 300$ K have a speed within ± 1 m/s of 400 m/s?

Translation A mole of nitrogen molecules has a mass of 28 g, so the mass m of a nitrogen molecule is

$$m = \frac{0.028 \text{ kg}}{1 \text{ mol}} \left(\frac{1 \text{ mol}}{6.02 \times 10^{23}} \right) = 4.65 \times 10^{-26} \text{ kg} \qquad (T7.7)$$

Let $v = 400$ m/s, and the range of acceptable speeds is $dv = 2$ m/s. Note that

$$v_P = \left(\frac{2k_B T}{m} \right)^{1/2} = \left[\frac{2(1.38 \times 10^{-23} \text{ J/K})(300 \text{ K})}{4.65 \times 10^{-28} \text{ kg}} \left(\frac{1 \text{ kg m}^2/\text{s}^2}{1 \text{ J}} \right) \right]^{1/2}$$

$$= 422 \text{ m/s} \qquad (T7.8)$$

Model The range dv in this case is pretty small compared to v_P, so we can use equation T7.4 to calculate the probability.

Solution We have

$$\left(\frac{v}{v_P} \right)^2 = \left(\frac{400 \text{ m/s}}{422 \text{ m/s}} \right)^2 = 0.899 \qquad \text{and}$$

$$\frac{dv}{v_P} = \frac{2 \text{ m/s}}{422 \text{ m/s}} = 4.74 \times 10^{-3} \qquad (T7.9)$$

So

$$\text{Pr(within } dv \text{ around } v) = \frac{4}{\pi^{1/2}} \left(\frac{v}{v_P} \right)^2 e^{-(v/v_P)^2} \frac{dv}{v_P}$$

$$= \frac{4}{\pi^{1/2}} (0.899) e^{-0.899} (4.74 \times 10^{-3})$$

$$= 3.91 \times 10^{-3} \qquad (T7.10)$$

Since there are 6.02×10^{23} molecules in our gas sample, we would expect about

$$(3.91 \times 10^{-3})(6.02 \times 10^{23}) = 2.36 \times 10^{21} \qquad \text{(T7.11)}$$

molecules to have speeds in this range.

Evaluation   Note that the probability comes out unitless, as it must. This also seems like a pretty reasonable fraction: *most* of the molecules will have speeds between roughly 0 and 800 m/s, and 2 m/s represents 2.5×10^{-3} of this range, so we might expect the probability to be about that order of magnitude.

Example T7.2

Problem   At $T = 300$ K, about what fraction of oxygen molecules in a sample of air will have a speed less than 200 m/s?

Model and Translation   Let $v_{\text{max}} \equiv 200$ m/s. We know from exercise T7X.1 that $v_P = 395$ m/s for oxygen at $T = 300$ K, so the range of interest here is *not* small compared to v_P. In such a case, we must integrate $\mathcal{D}(v)$ over the range in question:

$$\Pr(v < v_{\text{max}}) = \int_0^{v_{\text{max}}} \mathcal{D}(v)\,\frac{dv}{v_P} = \frac{4}{\pi^{1/2}} \int_0^{v_{\text{max}}} \left(\frac{v}{v_P}\right)^2 e^{-(v/v_P)^2}\,\frac{dv}{v_P}$$

$$= \frac{4}{\pi^{1/2}} \int_0^{x_{\text{max}}} x^2 e^{-x^2}\,dx \qquad \text{(T7.12a)}$$

where $\quad x \equiv \dfrac{v}{v_P} \quad$ and $\quad x_{\text{max}} = \dfrac{v_{\text{max}}}{v_P} = \dfrac{200\,\text{m/s}}{395\,\text{m/s}} = 0.51 \qquad \text{(T7.12b)}$

Solution   The integral over x, alas, cannot be expressed in terms of simple functions, but we can evaluate it numerically (by dividing the x axis into very small steps, multiplying the step width dx by the function height $x^2 e^{-x^2}$ for each step, and then summing to find the area under the curve). I have written a computer program called MBoltz that does exactly this (you can download MBoltz from the *Six Ideas* website). Figure T7.2 shows a screen shot of the program window when the program is correctly set up for this problem: I got the result displayed to the right of the integral by pressing the Evaluate

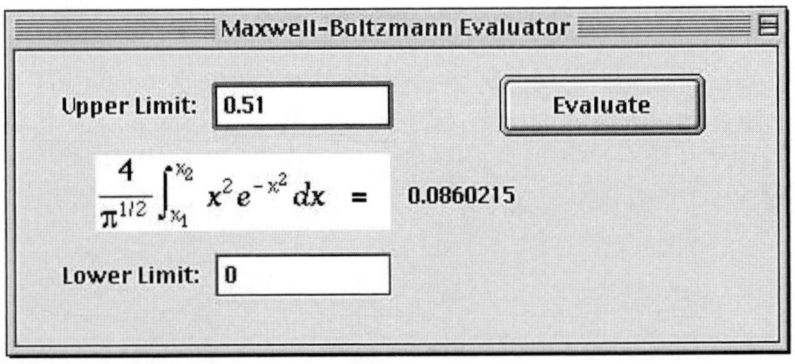

Figure T7.2
A screen shot of MBoltz set up to solve the problem discussed in example T7.2.

button. We see that the probability that an oxygen molecule has a speed in this range is about 0.086, so about 8.6% of oxygen molecules have a speed smaller than 200 m/s at this temperature.

Evaluation   This makes some sense: Figure T7.1 shows that $\mathscr{D}(v)$ is not very large throughout most of the range in question.

T7.2   Counting Velocity States

In this section, I will use the "quanton in a box" model discussed in unit Q to argue more rigorously that the assumption we made in section T7.1 about how molecular velocities are distributed is in fact correct. If you have not studied the quanton in a box model in your course so far, you can skip over this section, but if you *have* studied this model, reading this section will help you put the last section on a firmer foundation.

The number of quantum states consistent with a molecular speed in a given infinitesimal range

Consider first a gas molecule of mass m that can move freely in one dimension along the x axis between two walls a distance L apart. In unit Q, we saw that the molecule's wave function must go to zero at the walls, meaning that an integer number of half-wavelengths of that function must fit between the walls, as shown in figure T7.3. The wavelength λ of the molecule's wave function is linked to the molecule's x-momentum by the de Broglie relation $\lambda = h/p = h/|p_x|$, where h is Planck's constant and $p = |p_x|$ is the magnitude of its momentum. Therefore, requiring that we fit n half-wavelengths within the length L implies that

$$L = n\left(\frac{1}{2}\lambda\right) = n\left(\frac{h}{2|p_x|}\right) = n\left(\frac{h}{2m|v_x|}\right) \quad \Rightarrow \quad |v_x| = \frac{hn}{2mL} \quad (T7.13)$$

where n is a positive (nonzero) integer. We see that the molecule's x-velocity will be an integer multiple of $h/2mL$ (which is *very* small for any macroscopic sample of gas).

Exercise T7X.3

Show that if the gas is helium, and $L = 1$ mm, then $h/2mL = 5.0 \times 10^{-5}$ m/s, and the number is even smaller if L or m is larger. Note that the mass of 1 mol of helium is 4.0 g.

Now consider a molecule free to move in three dimensions in a cubic box whose sides have length L. Just as we had to fit an integer number of half-wavelengths between the walls limiting the molecule's motion along the x axis, we now have to fit (possibly different) integer numbers of half-wavelengths between the walls limiting its motion along the y and z axes. Therefore, all three of the molecule's velocity components will be integer

Figure T7.3

According to quantum mechanics, each molecule's wave function must fit an integer number of half-wavelengths between the container's boundaries at $x = 0$ and $x = L$.

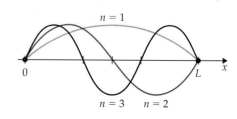

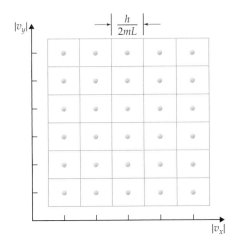

Figure T7.4
Each dot on this diagram represents a possible quantum state for a molecule moving in three dimensions (I have suppressed the $|v_z|$ dimension for clarity.) We can consider each state to occupy a "volume" of $(h/2mL)^3$ on such a graph.

multiples of $h/2mL$:

$$|v_x| = \frac{hn_x}{2mL} \qquad |v_y| = \frac{hn_y}{2mL} \qquad |v_z| = \frac{hn_z}{2mL} \qquad \text{(T7.14)}$$

where n_x, n_y, and n_z are independent positive integers. Each triplet of integers corresponds to a distinct molecular quantum state; so if we plot each state as a dot on a graph whose axes are $|v_x|$, $|v_y|$, and $|v_z|$, we see that the dots form a regularly spaced lattice, where each dot is separated from its nearest neighbor along each axis direction by $h/2mL$ (figure T7.4 shows the xy plane of this lattice). We can therefore think of each state as occupying the center of a tiny box of "volume" of $(h/2mL)^3$ on this graph.

The quantum states consistent with the molecule having a speed within a small range dv centered on some value v are those states that, when plotted as dots on a graph with axes $|v_x|$, $|v_y|$, and $|v_z|$, lie between the inner and outer surfaces of one-eighth of a spherical shell with thickness dv and radius v from the origin (one-eighth because the states all lie in the octant where $|v_x|$, $|v_y|$, and $|v_z|$ are all positive). The "volume" between the shell's surfaces on this graph is the shell's surface "area" $\frac{1}{8}4\pi v^2$ times its thickness dv (see figure T7.5),

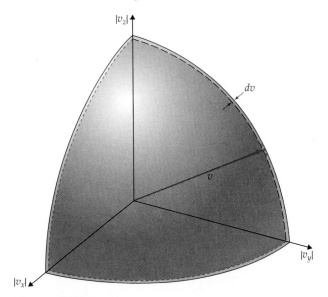

Figure T7.5
The quantum states corresponding to a molecular speed in a range dv centered on v are those that on a graph of v_x, v_y, and v_z lie between the two spherical surfaces shown.

which is proportional to $v^2\,dv$. Since each quantum state occupies the same very tiny volume on this graph, the number of quantum states within the shell will be proportional to the volume between the shell surfaces. Therefore the number of molecular quantum states within the specified range is indeed proportional to $v^2\,dv$, as claimed.

T7.3   The Average Energy of a Quantum System

An acorn-counting experiment

We can also use the Boltzmann factor to calculate the average energy that a quantum system will store when in contact with a reservoir at temperature T. An analogy may help you see how we can do this. Imagine that during a field biology lab, your lab group is asked to count the number of acorns that squirrels have stored in various dens. The results are shown in table T7.1, and figure T7.6 shows a histogram of the results.

To calculate the average number of acorns per den, you *could* type 5 zeros, 20 ones, 47 twos, 18 threes, 7 fours, and 3 fives into your calculator and then divide by 100. If you are a bit more clever and note that adding 20 ones is the same as adding $1 \cdot 20 = 20$, adding 47 twos is the same as adding $2 \cdot 47 = 74$, and so on, you can calculate the average more quickly as follows:

Table T7.1   Results of the acorn-counting experiment

Number of Acorns	Number of Dens with This Many Acorns
0	5
1	20
2	47
3	18
4	7
5	3
Total	100

$$\frac{\text{Average acorns}}{\text{Den}} = \frac{0 \cdot 5 + 1 \cdot 20 + 2 \cdot 47 + 3 \cdot 18 + 4 \cdot 7 + 5 \cdot 3}{100} = 1.96 \quad \text{(T7.15)}$$

Now note that 5/100 is the *probability* that a den has zero acorns, 20/100 is the *probability* that a den has one acorn, and so on. Since dividing the sum in the numerator in equation T7.15 by 100 is the same as dividing each term by 100 and then summing, we can also calculate the average in this way:

$$\frac{\text{Average acorns}}{\text{Den}}$$

$$= 0\left(\frac{5}{100}\right) + 1\left(\frac{20}{100}\right) + 2\left(\frac{47}{100}\right) + 3\left(\frac{18}{100}\right) + 4\left(\frac{7}{100}\right) + 5\left(\frac{3}{100}\right)$$

$$= 0 \cdot \Pr(0) + 1 \cdot \Pr(1) + 2 \cdot \Pr(2) + 3 \cdot \Pr(3) + 4 \cdot \Pr(4) + 5 \cdot \Pr(5) \quad \text{(T7.16)}$$

where $\Pr(n)$ is the probability that a den has n acorns.

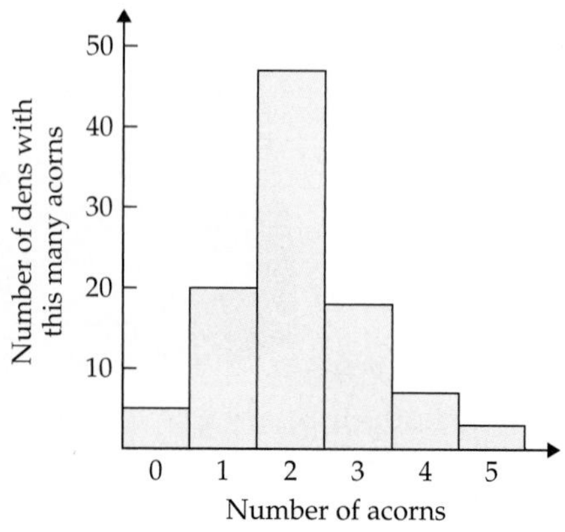

Figure T7.6
A histogram of the data in table T7.1.

Consider now a microscopic quantum system (such as an atom or a molecule) with energy levels E_0, E_1, E_2, and so on. By analogy with the above, we can compute the system's average energy in a given situation by computing the sum

How to find a quantum system's average energy

$$E_{avg} = E_0 \Pr(E_0) + E_1 \Pr(E_1) + E_2 \Pr(E_2) + \cdots = \sum_{\text{all } n} E_n \Pr(E_n) \quad (T7.17)$$

where $\Pr(E_n)$ is the probability that the system is in the nth quantum state. We know from chapter T6 that if the quantum system is in thermal contact with a reservoir at temperature T, its probability of being in a quantum state with energy E_n is simply

$$\Pr(E_n) = \frac{1}{Z} e^{-E_n/k_B T} \qquad \text{where} \qquad Z \equiv \sum_{\text{all } n} e^{-E_n/k_B T} \quad (T7.18)$$

Therefore, the average energy of a quantum system in contact with a reservoir is

$$E_{avg} = \sum E_n \left(\frac{e^{-E_n/k_B T}}{Z} \right) = \frac{\sum E_n e^{-E_n/k_B T}}{\sum e^{-E_n/k_B T}} \quad (T7.19)$$

The average energy for any quantum system in contact with a reservoir

Purpose: This equation describes how we can calculate the average energy E_{avg} of a quantum system in thermal contact with a reservoir at temperature T.

Symbols: E_n is the energy of the quantum system's nth quantum state, k_B is Boltzmann's constant, and $Z \equiv \sum e^{-E_n/k_B T}$. The sums are over *all* the system's states.

Limitations: This equation only works for a system in contact with something large enough to be considered a reservoir for the energies the system might typically have.

The average energy of a quantum system (qs) is interesting for a variety of reasons, but one of the most important is the following. Consider a macroscopic object at temperature T that is comprised of N_{qs} identical quantum systems. If N_{qs} is very large, we can imagine each quantum system to be in contact with a reservoir consisting of the remaining systems. Under these circumstances, the object's total thermal energy U is very nearly

Why the average energy is interesting

$$U = N_{qs} E_{avg} \quad (T7.20)$$

Therefore, knowing a quantum system's average energy at a given temperature T means that we can calculate the total thermal energy U of an object made up of such quantum systems without having to know anything about the system's multiplicity Ω. Since it is difficult to determine Ω for most systems other than an Einstein solid, this can be *very* useful.

Example T7.3

Problem   Imagine that a certain kind of molecule in a solution at room temperature can be in either of two quantum states with energies $E_0 = 0$ and $E_1 = \varepsilon$, respectively, where $\varepsilon = 0.02$ eV. What is the average energy of such a molecule? If the solution contains 10^{22} such molecules, what is the total thermal energy U stored by the excited molecules?

Model   The quantum system in this case is an individual molecule; the reservoir is the rest of the gas. Since each molecule is a quantum system, the number of molecules N is the same as the number of quantum systems N_{qs}.

Solution   Writing out the sums in equation T7.19 explicitly, we have in this case

$$E_{avg} = \frac{E_0 e^{-E_0/k_B T} + E_1 e^{-E_1/k_B T}}{e^{-E_0/k_B T} + e^{-E_1/k_B T}} = \frac{0 + \varepsilon e^{-\varepsilon/k_B T}}{e^{-0} + e^{-\varepsilon/k_B T}}$$

$$= \frac{\varepsilon e^{-\varepsilon/k_B T}}{1 + e^{-\varepsilon/k_B T}} = \varepsilon \left(\frac{1}{e^{\varepsilon/k_B T} + 1} \right) \tag{T7.21}$$

where in the last step I multiplied top and bottom by $e^{\varepsilon/k_B T}$. Since at room temperature $k_B T \approx 0.026$ eV, we see that in this case $\varepsilon/k_B T = (0.02\ \text{eV})/(0.026\ \text{eV}) = 0.77$, so

$$E_{avg} = \varepsilon \left(\frac{1}{e^{0.77} + 1} \right) = 0.32\varepsilon = 0.0063\ \text{eV} \tag{T7.22}$$

The total thermal energy U due to N such molecules being in the excited state is

$$U = NE_{avg} = 0.32N\varepsilon = 0.32(10^{22})(0.02\ \text{eV})\left(\frac{1.60 \times 10^{-19}\ \text{J}}{1\ \text{eV}} \right) = 10.2\ \text{J} \tag{T7.23}$$

Evaluation   Let's check equation T7.22 by using equation T7.18 to calculate the states' probabilities directly and then using equation T7.17 to find the average energy. The probability that the molecule is in the excited state is

$$\Pr(E_1) = \frac{e^{-\varepsilon/k_B T}}{Z} = \frac{e^{-\varepsilon/k_B T}}{e^0 + e^{-\varepsilon/k_B T}} = \frac{1}{e^{\varepsilon/k_B T} + 1} \tag{T7.24}$$

which is the same as the quantity in parentheses in equation T7.21. So we see in this case that the atom has a probability of 0.32 of being in the excited state and thus a probability of 0.68 of being in the ground state. The average energy is thus $0 \cdot 0.68 + \varepsilon(0.32) = 0.32\varepsilon$.

T7.4   Application to Einstein Solids

What if the number of quantum states is infinite?

Example T7.3 was easy because the atom had only two possible quantum states, but many quantum systems of interest have a large or even infinite number of states. For example, consider the problem of finding the average energy of a single quantum oscillator in thermal contact with the remaining oscillators in an Einstein solid. Each oscillator's quantum states have energies $E_n = n\varepsilon$, where ε is some quantum of energy and $n = 0, 1, 2$, etc. Equation T7.19 in this case becomes

$$E_{avg} = \frac{\sum \varepsilon n e^{-\varepsilon n/k_B T}}{\sum e^{-\varepsilon n/k_B T}} \tag{T7.25}$$

Since both sums here have an infinite number of terms, how can we calculate them?

The characteristic temperature T_ε

First note that since k_B has units of joules per kelvin, the quantity $T_\varepsilon \equiv \varepsilon/k_B$ has units of kelvins. This **characteristic temperature** is the temperature whose corresponding energy $k_B T_\varepsilon$ is equal to the characteristic energy ε

between quantum states of the system. Note that this characteristic temperature is *not* generally equal to the system's actual temperature T, but rather provides a system-specific benchmark against which T can be compared. The temperature ratio T/T_ε expresses the system's temperature T as a multiple of the system's characteristic temperature T_ε.

Now note that since $\varepsilon/k_B T = T_\varepsilon/T$, we can rewrite the Boltzmann factor $e^{-\varepsilon n/k_B T}$ appearing in equation T7.25 in the form $e^{-nT_\varepsilon/T}$. No matter how large T/T_ε is (and thus how small T_ε/T is), as we calculate the terms in the sums in equation T7.25, we will eventually get to the point where n is large enough that $-nT_\varepsilon/T$ is a large negative number and e^{-nT/T_ε} becomes very small. At this point, terms in each sum will become so small compared to that sum's first few terms that we can terminate the sum without much loss of accuracy.

Exercise T7X.4

For example, assume that $T/T_\varepsilon = 0.1$. Argue that one can get an excellent approximation for equation T7.25 by keeping only the first two terms in each sum.

If T/T_ε is large, though, we might have to sum *thousands* of terms before $e^{-nT_\varepsilon/T}$ becomes small. One way to calculate lengthy sums is to convert them to integrals. This is a powerful approach that yields excellent approximations when T/T_ε is very large, but it can involve some pretty gruesome mathematics. This is an approach generally discussed in statistical mechanics courses for upper-level physics majors.

However, if T/T_ε is not *too* large, one can use a simple computer program to do the sums instead. Such a program only does exactly what *you* would do if you were forced by some slave-driving professor to perform the sums. But not only does the computer not mind, it can also calculate thousands of terms in the blink of an eye. Such a program also gives good results for intermediate values of the temperature ratio T/T_ε, where a hand calculation is impractical but the integral is not a very good approximation to the actual discrete sum.

We can use a similar approach to evaluate equation T7.19 for other kinds of quantum systems. We can almost always write the energies of a system's quantum states in the form $E_n = \varepsilon f(n)$, where ε is some constant with units of energy and $f(n)$ is some unitless function of an integer quantum state index n. If we can, we can define a characteristic temperature $T_\varepsilon \equiv \varepsilon/k_B$ for the system, write the Boltzmann factor appearing in equation T7.19 as $e^{-f(n)T_\varepsilon/T}$, and sum terms until $f(n)$ becomes large enough so that the Boltzmann factor is small.

The program EBoltz (which you can download for free from the *Six Ideas* website) will calculate equation T7.19 using this approach. Figure T7.7 shows a screen shot of the program window. EBoltz calculates the expression displayed in the upper left corner of the window. Do not worry about what $g(n)$ is: we will set this function to be 1 for the time being. By entering a formula involving n in the text box in the upper right, you can specify E_n as being some characteristic energy ε times an arbitrary function of n. You can also set upper and lower limits on the value of n in the sum. When you press the Graph button, the program not only calculates the sums in equation T7.19 (which may involve thousands of terms before the terms are small enough to ignore) but also calculates these sums for many values of T (within a specified range) and plots $E_{avg}/k_B T$ versus $k_B T/\varepsilon = T/T_\varepsilon$ on the graph at the bottom. This is essentially a plot of E_{avg} (in units of $k_B T$) versus temperature (in units of T_ε).

We can adapt this approach for other kinds of systems

Using EBoltz to model the behavior of an Einstein solid

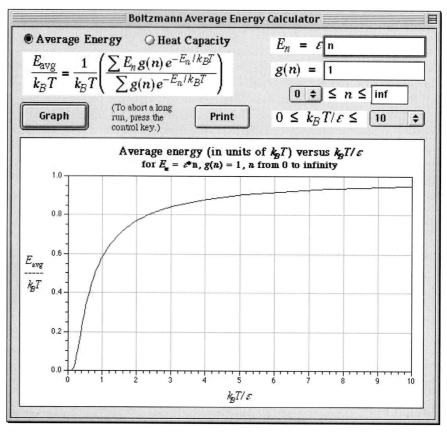

Figure T7.7

A screen shot showing EBoltz set up to calculate the average energy as a function of temperature for an oscillator in an Einstein solid.

Figure T7.7 shows the program configured to calculate E_{avg} for a single quantum harmonic oscillator in contact with a reservoir. We see from figure T7.7 that as the temperature ratio $T/T_{\varepsilon} = k_B T/\varepsilon$ becomes large, the oscillator has an average energy approximately equal to $k_B T$. This makes sense, because we saw in chapter T6 that an Einstein solid consisting of N atoms (and thus $N_{qs} = 3N$ oscillators) *should* have a total thermal energy of $U = N_{qs} k_B T = 3N k_B T$ at sufficiently high temperatures.

In practice, it's easier to measure an object's heat capacity dU/dT (the ratio of the tiny amount of energy dU we add and the tiny temperature change dT produced) than its total thermal energy U. Note that when our object consists of N_{qs} identical quantum systems, then

$$\frac{dU}{dT} = \frac{d(U/N_{qs})}{dT} N_{qs} = \frac{dE_{avg}}{dT} N_{qs} = \left(\frac{1}{k_B} \frac{dE_{avg}}{dT} \right) N_{qs} k_B \qquad \text{(T7.26)}$$

EBoltz can plot $k_B^{-1}(dE_{avg}/dT)$ as a function of temperature (see figure T7.8). The program evaluates $k_B^{-1}(dE_{avg}/dT)$ simply by evaluating E_{avg} at two different closely spaced values of T, finding the difference, and dividing by the difference in $k_B T$. Again, this is simple enough that you could do this yourself if you had the time.

Explaining results in table T2.2

We see from figure T7.8 that for temperatures $T > 3T_{\varepsilon}$, dU/dT for an Einstein solid is essentially 1.0 times $N_{qs} k_B = 3N k_B$. For lead, the effective value of ε is about 0.0057 eV, so the corresponding characteristic temperature is $T_{\varepsilon} \equiv \varepsilon / k_B = (0.0057\,\text{eV})/(8.62 \times 10^{-5}\,\text{eV/K}) = 66\,\text{K}$. Since room temperature (295 K) is about $4.6 T_{\varepsilon}$ in this case, we would expect from the graph that

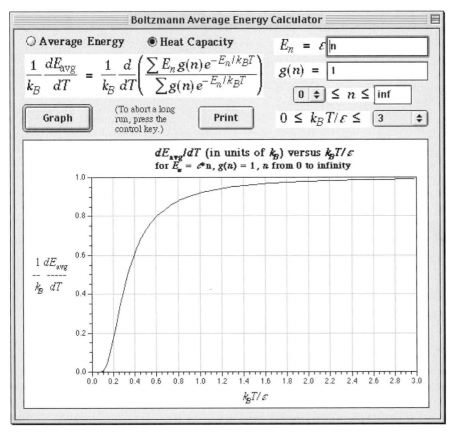

Figure T7.8
A screen shot of EBoltz set up to calculate the heat capacity as a function of temperature of an oscillator as a function of temperature.

dU/dT for lead would be almost exactly $3Nk_B$ at normal temperatures. The effective value for ε for aluminum, on the other hand, is about 0.026 eV, implying that aluminum's characteristic temperature is $T_\varepsilon = 300$ K. Since $T \approx T_\varepsilon$ at room temperature, the graph indicates that for aluminum at room temperature, we should expect $dU/dT \approx (0.92)(3Nk_B) = 2.76Nk_B$. The measured values displayed in table T2.2 for lead and aluminum are $3.18Nk_B$ and $2.92Nk_B$, respectively. These values are a bit higher than predicted because the atoms do not behave precisely as true harmonic oscillators. Note, however, that our model correctly predicts the *approximate* magnitudes and the fact that dU/dT is a bit smaller for aluminum than for lead (indeed, it gets the *ratio* of the two values of dU/dT almost exactly right).

Exercise T7X.5

Check that the ratio of the predicted values of dU/dT for aluminum and lead is almost exactly equal to the ratio of the measured values.

T7.5   Energy Storage in Gas Molecules

A diatomic gas molecule in principle can vibrate along its long axis. We can estimate the vibrational contribution to a gas molecule's average energy by modeling its vibrational aspect as a simple harmonic oscillator in contact

Why diatomic gas molecules do not vibrate

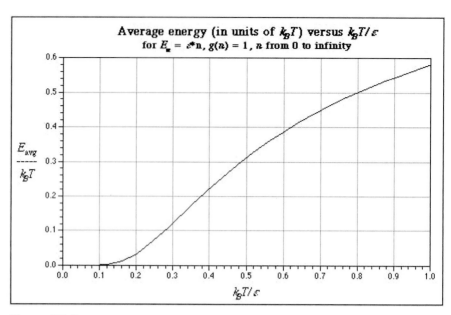

Figure T7.9

A graph showing the average energy per oscillator in an Einstein solid for low temperatures ($T/T_\varepsilon < 1$).

with a reservoir comprised of the rest of the gas molecules. The graphs shown in figures T7.7 and T7.8 therefore apply to this case as well. The difference is that the value of ε for a molecule such as H_2 is close to 0.44 eV, with a corresponding characteristic temperature $T_\varepsilon = 5100$ K. Even at 2000 K (where hydrogen begins to disassociate) the value of T/T_ε is less than 0.4. As the graphs in figures T7.9 and T7.8 show, the average vibrational energy that a hydrogen molecule stores and the amount that this vibrational energy contributes to the heat capacity dU/dT are essentially negligible at room temperature since $T \approx 0.06 T_\varepsilon$. Even at 2000 K, the contributions are small fractions of $k_B T$ and $N_{qs} k_B = N k_B$, respectively. The results are qualitatively similar for other diatomic molecules as well. This is why vibrational energy storage in diatomic molecules is essentially switched off until we reach temperatures much higher than room temperature.

We can understand this qualitatively by using the following model. At room temperature, the energy of the first excited vibrational quantum state of a hydrogen molecule is $\varepsilon \approx 0.44$ eV $\approx 17(0.026 \text{ eV}) = 17 k_B T$ above the ground state. For such a molecule to store any thermal energy at all, interactions with other molecules or with container walls must be able to bump the molecule up to at least this energy level. One can use the techniques of example T7.2 to show that the probability that a gas molecule will be moving fast enough to have a kinetic energy of $17 k_B T$ is about 2×10^{-7}. So even if a collision between a molecule and a wall in principle could convert the molecule's kinetic energy entirely to vibrational energy, in practice, the vast majority of molecules simply do not have nearly enough kinetic energy to get the molecule to even the first vibrational excited state, much less to higher states. Therefore, virtually all hydrogen molecules will remain in their vibrational ground state at room temperature, where their contribution to the thermal energy is zero.

Other diatomic molecules have marginally lower effective values of ε ($\varepsilon \approx 0.26$ eV for CO and $\varepsilon \approx 0.21$ eV for N_2, for example), but the temperature ratio T/T_ε at room temperature is still much too small for these molecules to store much vibrational energy.

Exercise T7X.6

Verify that the probability that a gas molecule has a kinetic energy above $17k_BT$ is about 2×10^{-7}. [*Hint:* First show that if a molecule's kinetic energy is K, then $K/k_BT = (v/v_P)^2$, where v is the molecule's speed and v_P is as defined in equation T7.4*b*.]

Rotational energy levels in diatomic molecules have a different spacing than harmonic oscillator energy levels, so the rotational average energy curves analogous to figures T7.7 and T7.8 are somewhat different. But *qualitatively*, they are pretty similar: they are very small compared to k_BT for temperatures below about $0.4T_\varepsilon$ and approach k_BT for temperatures larger than T_ε (see problems T7S.7 and T7S.8 for details) when $\varepsilon \equiv$ one-half of the energy difference between the ground and first excited rotational states. The main difference between the vibrational and rotational cases is that the value of ε for a typical diatomic molecule is in the ballpark of 0.0002 eV with a corresponding characteristic temperature of $T_\varepsilon \approx 2$ K. At room temperature, $\varepsilon \approx 0.08k_BT$, so molecules have plenty of kinetic energy at room temperature to bump molecules into excited rotational energy states, and the average rotational energy stored is essentially equal to k_BT. Only for temperatures significantly lower than about 2 K would rotational energy storage be switched off in the same way that vibrational energy storage is switched off at room temperature. Since all diatomic gases other than hydrogen liquefy at temperatures much larger than this, it is pretty safe to say that the rotational energy stored by a diatomic gas molecule at normal temperatures is $\approx k_BT$.

Rotational energy in diatomic molecules

Exercise T7X.7

Verify that $k_BT_\varepsilon \approx 0.0002$ eV at $T_\varepsilon = 2$ K.

For hydrogen, the effective value of ε is about 0.0076 eV, so T_ε is about 90 K. Therefore, we would expect rotational energy storage to be switched off at temperatures well below 90 K and switched on at temperatures well above this. This is *roughly* what we saw in figure T2.2. Predicting an exact transition temperature is complicated by a number of subtle quantum effects (see Schroeder, *Thermal Physics*, pp. 236–238 for a full discussion). A calculation taking careful account of these effects implies that $dU_{\rm rot}/dT$ should reach one-half of its full value of Nk_B at about 150 K, a result completely consistent with figure T2.2.

Explaining figure T2.2

Therefore, in broad strokes at least, we finally understand the features of this graph. At temperatures significantly below 150 K, hydrogen behaves as a monatomic gas because molecules rarely have enough energy to bump hydrogen molecules to their first excited rotational states, so rotational energy storage is switched off. At temperatures well above 150 K, this energy storage mode is fully switched on, so a hydrogen molecule stores an average rotational energy of k_BT in addition to its average kinetic energy of $(3/2)k_BT$. Because the energy difference between the ground state and first excited *vibrational* energy state is so high, though, vibrational energy storage in hydrogen is switched off until temperatures become quite high (it is only partially on when hydrogen begins to disassociate at 2000 K!).

The most important thing to carry away from this section is that the average energy $E_{\rm avg}$ and the specific heat $dE_{\rm avg}/dT$ of almost *any* quantum

The most crucial thing to understand

system in contact with a reservoir will behave qualitatively as shown in figures T7.7 and T7.8, respectively. We can almost always define a characteristic energy ε for the system that describes the spacing between its lowest energy levels and define a corresponding characteristic temperature $T_\varepsilon = \varepsilon / k_B$. If the system's temperature T is much less than T_ε, then random thermal interactions will generally not have enough energy to bump the system to any of its excited levels, so it will not be able to store a significant amount of energy, implying that the system is switched off as an energy storage mode. On the other hand, if $T \gg T_\varepsilon$, then random thermal interactions can easily bump the system into higher energy levels. At such temperatures, the quantum system's average energy is typically very nearly equal to $k_B T$ (though it can be $\frac{1}{2} k_B T$ in some cases), and we say that the system is "switched on."

The characteristic temperature T_ε therefore describes the critical temperature where a quantum system switches on. As the argument in the last paragraph makes clear, the very existence of a switching temperature depends on the ideas that the system's energy levels are quantized (in newtonian mechanics, a microscopic system can accept or give up arbitrarily small values of energy, so it is *always* fully switched on). The success of the quantum model in explaining the empirical results discussed in chapter T2 is strong evidence for its validity.

TWO-MINUTE PROBLEMS

T7T.1 Physically, why is a gas molecule's speed v more likely to be near v_P than near 0?
A. There are fewer quantum states with smaller speeds than larger speeds.
B. The Boltzmann factor is larger near $v = v_P$ than near $v = 0$.
C. We know that $\frac{1}{2}m[v^2]_{\text{avg}} = (3/2)k_B T$, and v_P is closer to $[v^2_{\text{avg}}]^{1/2}$ than 0 is.

T7T.2 Physically, why is a gas molecule's speed v more likely to be near v_P than near $2v_P$?
A. There are fewer quantum states for larger speeds than there are at speeds near v_P.
B. The Boltzmann factor is larger at v_P than at $2v_P$.
C. We know that $\frac{1}{2}m[v^2]_{\text{avg}} = (3/2)k_B T$, and v_P is closer to $[v^2_{\text{avg}}]^{1/2}$ than $2v_P$ is.

T7T.3 Imagine that the energy difference between adjacent energy levels in a harmonic oscillator is $\varepsilon = 0.005$ eV. At room temperature the value of the temperature ratio T/T_ε is
A. Quite a bit smaller than 1
B. About equal to 0.5
C. About equal to 1
D. About equal to 2
E. Quite a bit larger than 2
F. Not calculable since there is not enough information given to tell

T7T.4 In the situation described in problem T7T.3, we would expect the average energy of this oscillator when it is

in contact with a reservoir at room temperature to be
A. Negligible
B. Smaller than $\frac{1}{2}k_B T$
C. About equal to $\frac{1}{2}k_B T$
D. About equal to $k_B T$
E. Significantly larger than $k_B T$
F. Not calculable since there is not enough information to tell

T7T.5 Atoms can also store energy by bumping their electrons up to excited energy levels. For example, we could store 10.2 eV of energy in a hydrogen atom by bumping its single electron from the ground state up to its first excited energy level. Yet we do not consider this energy storage mode when calculating the thermal energy of gas molecules at room temperature. Why not?
A. Collisions with walls cannot cause the electron to change energy levels even in principle.
B. This kind of energy does not count as thermal energy.
C. Atoms can store energy in this way, but we ignore this mode because it is too complicated.
D. The electrons can get bumped up, but drop back to the ground state almost immediately, so on average, little energy is stored in this way.
E. At room temperature, T/T_ε is too low for this mode to be switched on.

T7T.6 Diatomic molecules can only vibrate along their length, but polyatomic molecules can often vibrate in several different ways. We can generally model each vibrational mode as if it were a simple harmonic oscillator independent of the others. Imagine that the difference between energy levels for one vibrational mode is $\varepsilon_1 = 0.01$ eV, while the same for another mode is $\varepsilon_2 = 0.04$ eV. For which vibrational mode will the average energy be larger at room temperature?

A. The first mode will have the larger average energy.
B. The second mode will have the larger average energy.
C. Both modes will store an average energy of about $k_B T$.
D. There is not enough information given to answer the question.

HOMEWORK PROBLEMS

Basic Skills

T7B.1 What is the probability that an oxygen molecule in a gas at room temperature will have a speed of $500 \text{ m/s} \pm 5 \text{ m/s}$?

T7B.2 (a) Show that for nitrogen at $0°C$, $v_P = 403$ m/s.
(b) What is the probability that a nitrogen molecule in a gas at this temperature will have a speed of $250 \text{ m/s} \pm 2 \text{ m/s}$?

T7B.3 Use the MBoltz program to find the probability that an oxygen molecule in a gas at room temperature will have a speed between 300 and 500 m/s.

T7B.4 Use the MBoltz program to find the probability that a gas molecule will have a speed greater than v_P.

T7B.5 Imagine that the energy difference between a simple quantum oscillator's energy levels is $\varepsilon = 0.035$ eV. What will be the average energy stored in this oscillator at room temperature? (*Hint:* You can read this from figure T7.7.)

T7B.6 Imagine that the energy difference between a simple quantum oscillator's energy levels is $\varepsilon = 0.035$ eV. At about what temperature will the average energy stored in this oscillator exceed $0.90 k_B T$?

Synthetic

T7S.1 Verify mathematically that the Maxwell-Boltzmann distribution function $\mathscr{D}(v)$ has its maximum value at $v = v_P$.

T7S.2 Imagine that in a certain physics experiment, we have a plasma of ionized oxygen atoms emerging from an opening in a furnace with a temperature of 3200 K. We then use a velocity selector (such as the one discussed in unit E) which passes only those ions within ± 0.01 km/s of 1.40 km/s. If we need about 2×10^{12} ions per second with the selected speed for our experiment, about how many ions per second must the furnace emit? (We can use this information to help determine what furnace to order.)

T7S.3 Water molecules evaporate from the surface of a puddle when random collisions give them a kinetic energy that exceeds the binding energy holding the molecules together, which is about 0.42 eV. As a first approximation, assume that the water molecules have speeds given by the Maxwell-Boltzmann distribution. About what fraction of molecules on the surface can evaporate if the puddle has a temperature of $20°C$? What is the fraction if the temperature is $50°C$? Use this to qualitatively explain why when you rub your hands under a bathroom hand dryer, most of the drying seems to happen during the last few seconds.

T7S.4 *Without* using MBoltz, estimate the fraction of molecules in a gas at temperature T whose speed is less than 0.1 times the most probable speed at that temperature. [*Hint:* It is *not* a good approximation to assume that this range of speeds is small enough that we can just multiply $\mathscr{D}(v)$ by $dv = 0.1 v_P$. Why not? So you will have to do an integral over $\mathscr{D}(v)$. But you can replace the exponential by something simpler over this range. You can *check* your work with MBoltz.]

T7S.5 Do the sums in equation T7.25 by hand for $T/T_\varepsilon = k_B T/\varepsilon = 0.2$ and $T/T_\varepsilon = 0.4$, and check that the EBoltz graph shown in figure T7.9 is correct at least at these two values. Do enough terms that your results are accurate to three decimal places.

T7S.6 (a) Imagine that a certain quantum system has quantum states whose energies are $E_n = \varepsilon n^2$, where $n = 0, 1, 2$, etc. Set up EBoltz for this situation, and determine the average energy of such a quantum system (as a multiple of $k_B T$) in the high-temperature limit. Submit a printout of the graph that was most helpful in answering this question.
(b) Imagine that a certain quantum system has quantum states whose energies are $E_n = \varepsilon \sqrt{n}$, where $n = 0, 1, 2$, etc. Set up EBoltz for this situation, and determine the average energy of such a quantum system (as a multiple of $k_B T$) in the high-temperature limit. Submit a printout of the graph that was most helpful in answering this question. [*Hint:* You can type "sqrt(n)" as the expression for n. Do not try for values of $T/T_\varepsilon = k_B T/\varepsilon$ much greater than 10, or you will wait for a very long time for the computer to finish its sums.]
(c) Compare your answers for the previous parts for the answers we found for the harmonic oscillator case ($E_n = \varepsilon n$). Do you see a pattern emerging? If so, what is it? Defend your response.

T7S.7 (a) Show that if a symmetric object has a moment of inertia of I and an angular momentum of L, we can write its rotational kinetic energy as

$$K^{\text{rot}} = \frac{L^2}{2I} \tag{T7.27}$$

(b) Model an oxygen molecule as two point particles of mass m connected by a massless rigid rod of length $D = 0.121$ nanometer (nm) (the approximate measured value of the O—O bond length). The mass of Avogadro's number of oxygen *atoms* is 16 g. Calculate the moment of inertia I for this molecule for rotations around its center of mass (perpendicular to the rod).

(c) The earliest forms of quantum theory proposed that all angular momenta were quantized in units of $\hbar \equiv h/2\pi$. If this is true, then the rotational energy levels of any diatomic molecule rotating around an axis perpendicular to its length would be

$$E_n = \frac{(\hbar n)^2}{2I} = \varepsilon n^2 \quad \text{where} \quad \varepsilon \equiv \frac{\hbar^2}{2I}$$

and $j = 0, 1, 2,$ etc. $\tag{T7.28}$

Show that for oxygen the value of ε is 0.0002 eV to one significant digit.

(d) An oxygen molecule also might in principle rotate around its long axis. Model each oxygen atom as having a spherical nucleus of radius 3.0 femtometers (fm) surrounded by a spherical electron cloud of radius 0.06 nm. As a rough approximation, imagine each spherical cloud to have uniform density. Note that electrons are roughly 2000 times less massive than a proton or neutron. Argue that the moment of inertia for an oxygen molecule rotating around its long axis is about a factor of 10,000 times smaller than when it is rotating around an axis perpendicular to its long axis.

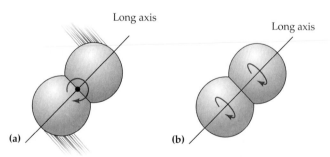

(a) Rotation of an oxygen molecule around an axis perpendicular to its long axis. (b) Rotation of the same around the long axis.

T7S.8 Modern quantum mechanics asserts that the rotational energy levels of a diatomic molecule rotating

around an axis perpendicular to its length are

$$E_n = \varepsilon n(n+1) \quad \text{where} \quad \varepsilon \equiv \frac{\hbar^2}{2I}$$

and $j = 0, 1, 2,$ etc. $\tag{T7.29}$

Moreover, there turn out to be $2n + 1$ distinct quantum states that have the same energy E_n (corresponding to different possible orientations of the molecule's angular momentum vector in space). Now, the sums appearing in equation T7.19 for the average energy are sums over *all quantum states*. Since all states with the same n have the same energy E_n and thus the same Boltzmann factor $e^{-E_n/k_B T}$, they will contribute equally in both sums. So we can convert these sums to sums over j by multiplying each term in the sum by $2n + 1$:

$$E_{\text{avg}} = \frac{\sum_{n=0}^{\infty} E_n g(n) e^{-E_n/k_B T}}{\sum_{n=0}^{\infty} g(n) e^{-E_n/k_B T}}$$

where $E_n = \varepsilon n(n+1) \quad \varepsilon \equiv \frac{\hbar^2}{2I}$

and $g(n) = 2n + 1$ $\tag{T7.30}$

(a) Set up EBoltz to handle this situation. What are the values of the average rotational energy per molecule in the high-temperature limit? What is the contribution of rotational energy to the heat capacity dU/dT in the same limit? At what values of $T/T_\varepsilon = k_B T/\varepsilon$ do these quantities achieve one-half of their high-temperature values? Submit printouts supporting your claims.

(b) Considering your graphs for part (a) and the answers to parts (b) and (d) of problem T7S.7, why are rotations of a diatomic molecule around an axis perpendicular to its length switched on at room temperature, but rotations around the long axis are switched off?

T7S.9 According to equation T7.3, the probability that a molecule has a speed lying within a range dv centered on a certain speed v is

$$\text{Pr(range } dv \text{ centered on } v) = C\left(\frac{v}{v_P}\right)^2 e^{-(v/v_P)^2} \frac{dv}{v_P}$$

where $v_P \equiv \sqrt{\frac{2k_B T}{m}}$ $\tag{T7.31}$

Since the probability must be 1 that a gas molecule has some speed (any speed) between zero and infinity, we must have

$$1 = C\int_0^\infty \left(\frac{v}{v_P}\right)^2 e^{-(v/v_P)^2} \frac{dv}{v_P} = C\int_0^\infty x^2 e^{-x^2}\, dx$$

where $x \equiv \frac{v}{v_P}$ $\tag{T7.32}$

Look up the last integral in a table of integrals and determine the value of C.

T7S.10 In this problem, we will use the methods of section T7.5 to compute the average kinetic energy of a gas molecule in a container, considering each gas molecule as a quantum system in contact with the reservoir consisting of the rest of the gas. Consider first a gas molecule constrained to move in one dimension along the x axis between two container walls a distance L apart. We saw in section T7.2 that such a molecule's x-velocity must be an integer multiple of $h/2mL$ (see equation T7.13). The contribution that its motion along the x axis makes to its total kinetic energy K is therefore

$$\frac{1}{2}mv_x^2 = \frac{m}{2}\left(\frac{hn}{2mL}\right)^2 = \frac{h^2 n^2}{8mL^2}$$

where $n = 1, 2, 3, \ldots$ (T7.33)

As discussed in chapter T4, a molecule's ground state energy does not count as part of its thermal energy, since it cannot be changed by thermal interactions. We can therefore write the *thermal* part of this contribution to the molecule's kinetic energy as

$$E_{\text{th},n} = \varepsilon n^2 - \varepsilon = \varepsilon(n^2 - 1) \quad \text{where} \quad \varepsilon = \frac{h^2}{8mL^2}$$

and $n = 1, 2, 3, \ldots$ (T7.34)

Note that $E_{\text{th}} = 0$ when the molecule is in the ground state $n = 1$.

(a) Use EBoltz to evaluate the average energy of this contribution to the molecule's kinetic energy. (Note that you will have to change the lower limit of the sum from 0 to 1.) Submit printouts showing graphs over the temperature ranges $0 \le T \le 10T_\varepsilon$ and $0 \le T \le 1000T_\varepsilon$, where $T_\varepsilon \equiv \varepsilon/k_B$. What is the apparent value of this average energy as the temperature becomes extremely large?

(b) Imagine that we are talking about molecules of helium confined in a container whose distance L between the walls is only 1 mm. Calculate the value of T_ε in this case.

(c) Will making the container wall separation L larger make the value of T_ε larger or smaller? Explain.

(d) Will this energy storage mode be switched on or off at normal temperatures? Are we *ever* likely to see this quantum system in the other state (i.e., switched on if it is normally switched off or switched off if it is normally switched on)? Explain.

(e) The energies associated with motion along the y and z directions are completely independent (see equation T7.14) but have the same form as equation T7.34 (though the distances L between the container walls in those directions might be different). What do your answers to the previous parts imply that the total average kinetic energy of a gas molecule in a container of any reasonable size will be when it is in thermal contact

with a reservoir at any reasonable temperature? Explain.

Rich-Context

T7R.1 (a) The temperature of the upper atmosphere is actually fairly high, about 1000 K. A molecule in the upper atmosphere can escape if it has a speed that exceeds the earth's escape speed of 11.2 km/s. What is the probability that an N_2 molecule has such a speed? What is the probability that an H_2 molecule has such a speed? Why do you think earth's atmosphere contains nitrogen but not a significant amount of hydrogen?

(b) The escape speed at the surface of the moon is 2.4 km/s, and its surface temperature can exceed 375 K. Why does the moon have no significant atmosphere? (*Hint:* Think about what might be the most massive, reasonably common, naturally occurring molecules in the earth's atmosphere.)

T7R.2 (a) Consider two identical imaginary boxes of air with volume V, one at some altitude z and the other at sea level (taken to be $z = 0$), with the change in altitude much larger than the box size. The quantum states available to molecules in the higher box are essentially the same as those available to molecules in the lower one *except* that the energy of each quantum state of a molecule of mass m in the upper box is mgz higher than the corresponding state in the lower box, because of the gravitational potential energy involved. Now imagine that we connect these boxes with a hose, so that molecules can flow freely between the boxes. Use the Boltzmann factor to argue that in equilibrium, if the gas has the same temperature in both boxes, the ratio of the numbers of molecules in the two boxes will be

$$\frac{N(z)}{N(0)} = e^{-mgz/k_B T}$$ (T7.35)

where $N(z)$ is the number of molecules in the upper box and $N(0)$ is the number in the lower box.

(b) Argue that the ratio of the gas *pressures* P in each box is the same

$$\frac{P(z)}{P(0)} = e^{-mgz/k_B T}$$ (T7.36)

(c) We can model the earth's atmosphere as a series of vertically stacked and connected boxes of gas, so the last equation should apply to the earth's atmosphere to the extent that we can consider the earth's atmosphere to have a temperature independent of z (this is an **isothermal atmosphere model**). Use this model to predict the approximate air pressure at the top of Mount Everest (altitude of 8848 m).

(d) What is the approximate *density* of air at Coors Stadium in Denver, CO (at an altitude of about 1,610 m), as a fraction of the density at sea level? Do you think that this might significantly affect how far a well-hit baseball might travel in that stadium compared to how far it might travel in Yankee Stadium in New York City (which is barely above sea level)?

Closing comments: How good is this model? Up to an altitude of about 11 km, the temperature of the earth's atmosphere actually falls with increasing height, so we might expect the pressure to decrease actually a bit more rapidly with height than the isothermal model would suggest. Still, because we are measuring temperatures on the absolute scale, even an extreme temperature such as $-60°C = 213$ K is only about 25% smaller than room temperature (298 K), so the isothermal model should be a reasonable first approximation. Measurements show that the pressure of the earth's atmosphere as a function of height is pretty close to an exponential, but does indeed decrease a bit faster than the isothermal model would predict, and is very roughly 20% lower than the isothermal prediction at $z = 10$ km.

What is the air pressure at the top of Mount Everest?

Advanced

T7A.1 According to the argument in section T7.3, the average speed of a molecule in a gas is

$$v_{avg} = \sum_{\text{all speed ranges } i} v_i \, \Pr(v_i)$$

$$= \sum_{\text{all speed ranges}} v_i \, \mathscr{D}(v_i) \, dv \qquad (T7.37)$$

(a) Convert this sum to an integral to get

$$v_{avg} = \frac{4v_P}{\pi^{1/2}} \int_0^\infty x^3 e^{-x^2} \, dx$$

where $x = \dfrac{v}{v_P}$ \qquad (T7.38)

(b) Look up this integral in a table of integrals, and use the result to show that

$$v_{avg} = \sqrt{\frac{8k_B T}{\pi m}} \qquad (T7.39)$$

as claimed in equation T7.5.

T7A.2 Complicated quantum systems can store energy in a number of different ways. In this chapter, we have analyzed each energy storage mode *independently*, completely ignoring the other ways that the system can store energy! Is this legal? In this problem you will prove mathematically that it is. Consider as an example a molecule that can store energy in the form of both vibrational energy and rotational energy. Assume the molecule's vibrational energy levels E_{vib} depend on an integer n, while its rotational energy levels E_{rot} depend on some independent integer j, so that the molecule's total energy is $E_{tot} = E_{vib}(n) + E_{rot}(j)$. Note that the denominator Z in the expression for the average energy in equation T7.19 can be written

$$Z = \sum_{\text{all states}} e^{-E_{tot}/k_B T} = \sum_{\text{all } n} \sum_{\text{all } j} e^{-[E_{vib}(n)+E_{rot}(j)]/k_B T}$$

$$= \sum_{\text{all } n} \sum_{\text{all } j} e^{-E_{vib}(n)/k_B T} e^{-E_{rot}(j)/k_B T} \qquad (T7.40)$$

We can add the terms in any order that we like, so let us sum over j first. The value of $e^{-E_{vib}(n)/k_B T}$ is the same for all terms in the sum over j for any given value of n, so we can pull this factor out in front of the sum over j as follows:

$$Z = \sum_{\text{all } n} \left[e^{-E_{vib}(n)/k_B T} \left(\sum_{\text{all } j} e^{-E_{rot}(j)/k_B T} \right) \right]$$

$$= \sum_{\text{all } n} \left[e^{-E_{vib}(n)/k_B T} (Z_{rot}) \right] \qquad (T7.41)$$

where Z_{rot} is the value of Z we would have gotten if we had thought the molecule only had rotational energy. Since Z_{rot} has the same numerical value for all the vibrational states, we can factor it out of the sum over n, yielding

$$Z = Z_{rot} \sum_{\text{all } n} e^{-E_{vib}(n)/k_B T} = Z_{rot} Z_{vib} \qquad (T7.42)$$

where Z_{vib} is the value of Z we would have gotten if we had thought the molecule only had vibrational energy.

(a) In a similar way, show that

$$\sum_{\text{all states}} E_{vib}(n) e^{-[E_{vib}(n)+E_{rot}(j)]/k_B T}$$

$$= Z_{rot} \sum_{\text{all } n} E_{vib}(n) e^{-E_{vib}(n)/k_B T} \qquad (T7.43a)$$

and $\displaystyle\sum_{\text{all states}} E_{\text{rot}}(n)e^{-[E_{\text{vib}}(n)+E_{\text{rot}}(j)]/k_B T}$

$$= Z_{\text{vib}}\sum_{\text{all } n} E_{\text{rot}}(n)e^{-E_{\text{rot}}(n)/k_B T} \quad \text{(T7.43b)}$$

Be sure to carefully explain your reasoning.

(b) Combine these results with the result given in equation T7.42 to prove that

$$E_{\text{avg}} = E_{\text{vib,avg}} + E_{\text{rot,avg}} \quad \text{(T7.44)}$$

as claimed.

ANSWERS TO EXERCISES

T7X.1 According to equation T7.4b, we have

$$v_P = \sqrt{\frac{2k_B T}{m}} = \sqrt{\frac{2k_B T}{M_A/N_A}}$$

$$= \sqrt{\frac{2(1.38 \times 10^{-23} \text{ J/K})(300 \text{ K})}{(0.032 \text{ kg})/(6.02 \times 10^{23})}\left(\frac{1 \text{ kg m}^2/\text{s}^2}{1 \text{ J}}\right)}$$

$$= 395 \text{ m/s} \quad \text{(T7.45)}$$

T7X.2 We have

$$\frac{1}{2}m[v^2]_{\text{avg}} = \frac{3}{2}k_B T \quad \Rightarrow \quad [v^2]_{\text{avg}} = \frac{3k_B T}{m}$$

$$\Rightarrow \quad v_{\text{rms}} \equiv \sqrt{[v^2]_{\text{avg}}} = \sqrt{\frac{3k_B T}{m}}$$

$$\text{(T7.46)}$$

as stated.

T7X.3 The molecule's mass m is its mass per mole M_A divided by Avogadro's number N_A. Plugging in the numbers, we therefore find

$$\frac{h}{2mL} = \frac{hN_A}{2M_A L}$$

$$= \frac{(6.63 \times 10^{-34} \text{ J s})(6.02 \times 10^{23})}{2(0.004 \text{ kg})(0.001 \text{ m})}\left(\frac{1 \text{ kg m}^2/\text{s}^2}{1 \text{ J}}\right)$$

$$= 5.0 \times 10^{-5} \text{ m/s} \quad \text{(T7.47)}$$

T7X.4 When $T/T_\varepsilon = 0.1$, the values of $e^{-nT_\varepsilon/T} = e^{-10n}$ for the first few values of n are

$$e^0 = 1 \quad e^{-10} = 4.5 \times 10^{-5} \quad e^{-20} = 2.1 \times 10^{-9}$$
$$e^{-30} = 9.4 \times 10^{-14} \quad \text{(T7.48)}$$

and so on. So if we omit all terms in both sums beyond the first two, we will get a result accurate to within a few parts in a billion!

T7X.5 The predicted ratio is about 0.92. The actual ratio is

$$\frac{2.92Nk_B}{3.18Nk_B} = 0.918 \quad \text{(T7.49)}$$

T7X.6 According to equation T7.4b, $v_P^2 \equiv m/2k_B T$. This means that

$$\frac{K}{k_B T} = \frac{\frac{1}{2}mv^2}{k_B T} = \frac{m}{2k_B T}v^2 = \frac{v^2}{v_P^2} = x^2 \quad \text{(T7.50)}$$

where $x \equiv v/v_P$, as defined in example T7.2. So if $K > 17k_B T$, this means that $x > \sqrt{17} = 4.12$. To find the probability that a molecule will have a speed such that $x > 4.12$, we have to integrate the Maxwell-Boltzmann distribution function from $x = 4.12$ to infinity. Plugging in these numbers as the upper and lower limits in the program MBoltz returns 2.0149×10^{-7} (as shown below).

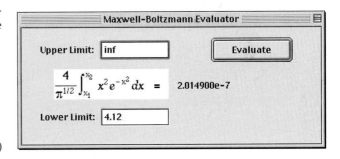

T7X.7 Since $k_B T_{\text{room}} \approx 0.026 \text{ eV}$,

$$k_B T_\varepsilon = (k_B T_{\text{room}})\left(\frac{T_\varepsilon}{T_{\text{room}}}\right) = (0.026 \text{ eV})\left(\frac{2 \text{ K}}{300 \text{ K}}\right)$$

$$= 1.7 \times 10^{-4} \text{ eV} \quad \text{(T7.51)}$$

If we round to one significant digit, this is the same as 0.0002 eV.

T8

Calculating Entropy Changes

Chapter Overview

Introduction

This chapter develops tools for computing entropy changes in processes involving volume changes and in cases where we do not know an object's multiplicity. These tools will be very useful in chapter T9.

Section T8.1:   The Entropy of a Monatomic Gas

One can very roughly estimate the multiplicity of an ideal monatomic gas by treating each momentum component of each molecule as a separate quantum system, estimating the number of states available to that system (i.e., the multiplicity of that system), and calculating the product of all these multiplicities to find the multiplicity of the gas as a whole. After correcting for the indistinguishability of the gas molecules, we find that

$$\Omega(U, V, N) \approx \frac{1}{N!}\left(\frac{8mV^{2/3}bU}{3Nh^2}\right)^{3N/2} \tag{T8.7}$$

$$S(U, V, N) = \frac{3}{2}Nk_B \ln\left(\frac{8mbV^{2/3}U}{3h^2 N}\right) - k_B \ln(N!) \tag{T8.8}$$

    Purpose: This equation expresses the multiplicity Ω and entropy S of an ideal monatomic gas as a function of its thermal energy U, volume V, and number of molecules N.

    Symbols: k_B is Boltzmann's constant, h is Planck's constant, m is the mass of a molecule, and b is some unitless constant (the approximation is best if we choose $b = \frac{1}{2}\pi e$).

    Limitations: This equation applies only to a monatomic gas, and only in the limit that both N and $n = (8mbV^{2/3}U/3Nh^2)^{1/2}$ are large. (Under normal conditions, monatomic gases *easily* satisfy these limitations.)

Section T8.2:   Entropy Depends on Volume

One can see from this formula that a gas's entropy increases as its volume V increases. This is so because increasing the volume available to the gas lowers the energies of all its quantum states, increasing the number of possibilities for distributing the gas's total energy among those states.

    Equation T8.7 in fact implies that $\Omega \propto V^N$ (a result that is true for *all* types of gases). This implies that the number of quantum states available to each molecule in the gas increases in proportion to V. Naively, one might think that this is so because the number of possible positions available to a molecule increases in proportion to V. While this idea is not quite correct (the multiplicity counts *energy* states, not position states), it does yield the right answer and is a useful way to remember the result.

Section T8.3: A General Expression for Entropy Changes

The definition of temperature implies that $dS = T^{-1}dU$ during an infinitesimal constant-volume process, but this equation does not work for processes involving volume changes, because, as we've seen, a gas's entropy S can change if its volume V changes even if its thermal energy U does not. The first step toward a more general formula for computing entropy changes is to recognize that, according to equation T8.8, $dS = 0$ *for a gas undergoing a* **quasistatic** *adiabatic volume change* (a result that actually applies to not only all kinds of gases but nongases as well). The difference between an adiabatic process and a more general process is that in the general process, some heat dQ and/or work dW_{other} not related to the quasistatic volume change flows into or out of the system: it is *this* energy that causes the entropy change. If $dW_{other} = 0$, then this extra energy is entirely heat, and

$$dS = \frac{dQ}{T} \qquad \text{(T8.17)}$$

Purpose: This equation tells us how to calculate a system's entropy change dS during an infinitesimal process.
  Symbols: dQ is the infinitesimal heat flowing into the system, and T is the system's temperature.
  Limitations: The value of dQ must be small enough that T does not change significantly during the process. The number of molecules N in the system must be constant, its volume V must change quasistatically, and there must be no work involved other than that due to the volume change.

Section T8.4: Constant-Temperature Processes

To compute ΔS for any finite process, we have to integrate equation T8.17: $\Delta S = \int T^{-1}dQ$. This is easy to do for an object if either its temperature or its specific heat is approximately constant during the process. In the first case, we can pull T out in front of the integral to get

$$\Delta S = \frac{Q}{T} \qquad \text{if } T \approx \text{constant during a process} \qquad \text{(T8.19)}$$

This might be the case if (1) the object in question is large enough to be considered reservoir, (2) it is in good thermal contact with a reservoir, or (3) it is undergoing a **phase change** (then $Q = \pm mL$, where m is the object's mass and L its **latent heat**).

Section T8.5: Handling Changing Temperatures

If an object's specific heat c is approximately constant during the process and its volume does not change significantly, then we can substitute $dQ = mc\,dT$ into the integral (where m is the object's mass), pull mc out of the integral, and integrate with respect to T to get

$$\Delta S = mc \ln \frac{T_f}{T_i} \qquad \text{if no work is done and } c \approx \text{constant} \qquad \text{(T8.30)}$$

Section T8.6: Non-quasistatic Processes

A system's entropy by definition depends only on its macrostate. Therefore, the *change* in a system's entropy depends only on the system's initial and final macrostates, and not at all on the process that gets it from one to the other. This means we can calculate the entropy change involved in *any* process, even if the process does not fit into one of the categories above, if we can find a **replacement process** involving a sequence of processes that *do* all fit into the categories above and take us from the same initial macrostate to the same final macrostate. The entropy change we compute for the replacement process will be the same as that for the original process.

T8.1   The Entropy of a Monatomic Gas

Estimating the multiplicity of
an ideal gas

Finding an *accurate* formula for the multiplicity of an ideal monatomic gas
is much more difficult than finding a formula for the Einstein solid, but we
can make a pretty good estimate using the following simple model. In section T7.2, we saw that when a monatomic gas molecule of mass m is confined to a cubic box with sides of length L, its total energy is equal to its
kinetic energy

$$K = \frac{1}{2}mv^2 = \frac{m}{2}\left(v_x^2 + v_y^2 + v_z^2\right) = \frac{h^2 n_x^2}{8mL^2} + \frac{h^2 n_y^2}{8mL^2} + \frac{h^2 n_z^2}{8mL^2} \quad \text{(T8.1)}$$

where n_x, n_y, and n_z are *independent* positive integers. Because these integers
are independent, we can model each molecule as a set of three identical but
independent quantum systems whose energy levels are

$$E_n = \varepsilon n^2 \quad \text{where} \quad \varepsilon \equiv \frac{h^2}{8mL^2} \quad \text{and} \quad n = 1, 2, 3, \ldots \quad \text{(T8.2)}$$

The *average* energy of each such system will be $E_{\text{avg}} = U/3N$, where U is the
gas's total thermal energy and N is the number of molecules.

Now, we know that the probability that the system will be in a given
quantum state decreases exponentially as the state's energy increases:
$\Pr(E_n) \propto e^{-E_n/k_B T}$ (see figure T8.1a). The probabilities of states whose energies are much larger than E_{avg} are so low that we can consider them to be
zero for practical purposes. So for the sake of argument, let us *pretend* that all
states whose energies are smaller than some multiple bE_{avg} of E_{avg} are
equally probable, and states with energies above that limit are impossible:
this amounts to replacing the actual probability distribution shown in figure T8.1a by the simplified distribution shown in figure T8.1b. (We have no
way to choose a specific value for b at this point, but it will turn out that the
value of b will not matter much for our purposes.)

Therefore, according to our model, the quantum states that are "accessible" to each system are those whose energies are less than bE_{avg}. The number
of such states is essentially the same as the value of n at the energy bE_{avg},

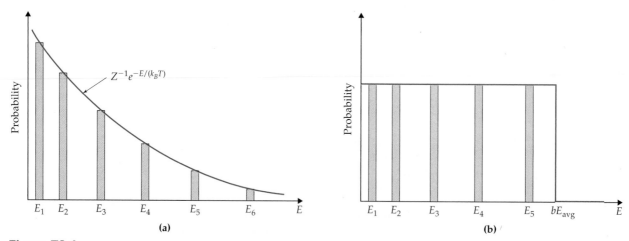

Figure T8.1

(a) This graph shows the exponentially decaying probability of energy states of a one-gas-molecule quantum system.
(b) In our simplified model, we consider states whose energies are beyond some multiple of the system's average
energy to have probabilities small enough to ignore, and we treat the rest as having equal probabilities.

which is given by

$$\varepsilon n^2 = b E_{avg} = \frac{bU}{3N} \quad \Rightarrow \quad n = \left(\frac{bU}{3N\varepsilon}\right)^{1/2} \qquad \text{(T8.3)}$$

Since these states are equally probable in our model, this is essentially the multiplicity of each of our $3N$ quantum systems. The multiplicity of a combined system is the product of the multiplicities of its parts, so the total multiplicity of our gas should therefore be

$$\Omega \approx \left[\left(\frac{bU}{3N\varepsilon}\right)^{1/2}\right]^{3N} = \left(\frac{bU}{3N\varepsilon}\right)^{3N/2} \qquad \text{(T8.4)}$$

(For a newtonian model leading to the same equation, see problem T8S.3.)

This result would be correct if the molecules in the gas were all distinguishable. In this model, we specify a microstate of the whole gas by listing a triplet of integers n_x, n_y, and n_z for each molecule. But since the molecules are all identical and free to move about, we really cannot keep track of which molecule belongs to which triplet. Since we cannot tell the molecules apart, if we were to take a microstate and shuffle the triplets around among the molecules, we would get a microstate that is operationally indistinguishable from the first. The rule in quantum mechanics in such circumstances is that indistinguishable microstates do not count as separate states. Since there are $N!$ ways to shuffle N molecules around without changing the set of integers, equation T8.4 overcounts the number of truly distinguishable states by a factor of $N!$; so a better estimate of the multiplicity is

Allowing for the indistinguishability of gas molecules

$$\Omega \approx \frac{1}{N!}\left(\frac{bU}{3N\varepsilon}\right)^{3N/2} \qquad \text{(T8.5)}$$

Now, this model looks pretty bogus: what makes us think we can get away with replacing the probability distribution in figure T8.1a with the one shown in figure T8.1b? It turns out that calculating the multiplicity of a monatomic ideal gas by directly counting states (with the help of some higher mathematics) in fact yields

Assessing the model's validity

$$\Omega \approx \frac{1}{N!}\left(\frac{\pi e U}{6N\varepsilon}\right)^{3N/2} \qquad \text{(T8.6)}$$

in the limit that $n = (bU/3N\varepsilon)^{1/2} \gg 1$ and $N \gg 1$. So our bogus estimate turns out to be basically correct in this limit if we choose $b = \frac{1}{2}\pi e = 4.27$. Moreover, one can show (see problem T8R.1) that if we apply the same approach to the Einstein solid, we find that $\Omega \approx (bU/3N\varepsilon)^{3N}$, which yields the same estimate found in problem T6A.1 (using a completely different method) if we set $b = e$. So this model must capture something valid about the multiplicity, at least in the large-n, large-N limit.

Exercise T8X.1

The number of states n that are accessible to a quantum system should be an integer, but we are approximating it by a function $(bU/3N\varepsilon)^{1/2}$ that is not necessarily an integer. Explain why this means that equation T8.3 will be a good approximation only when $n \gg 1$.

Using equation T8.5, the definition of ε, the fact that the gas's volume is $V = L^3$, and the definition of entropy, you can show that the multiplicity and entropy of an ideal monatomic gas in terms of the basic variables U, V, and N are

The entropy of an ideal
monatomic gas

$$\Omega(U, V, N) \approx \frac{1}{N!} \left(\frac{8mV^{2/3}bU}{3Nh^2} \right)^{3N/2} \tag{T8.7}$$

$$S(U, V, N) = \frac{3}{2} Nk_B \ln \left(\frac{8mb\,V^{2/3}U}{3h^2 N} \right) - k_B \ln(N!) \tag{T8.8}$$

Purpose: This equation expresses the multiplicity Ω and entropy S of an ideal monatomic gas as a function of its thermal energy U, volume V, and number of molecules N.
Symbols: k_B is Boltzmann's constant, h is Planck's constant, m is the mass of a molecule, and b is some unitless constant (the approximation is best if we choose $b = \frac{1}{2}\pi e$).
Limitations: This equation applies only to a monatomic gas, and only in the limit that both N and $n = (8mb\,V^{2/3}U/3Nh^2)^{1/2}$ are large. (Under normal conditions, monatomic gases *easily* satisfy these limitations; see problem T8S.2.)

Exercise T8X.2

Show that equations T8.7 and T8.8 follow from equation T8.5, $V = L^3$, $\varepsilon \equiv h^2/8mL^2$, and the definition of entropy $S \equiv k_B \ln \Omega$.

Example T8.1

Problem   Use equation T8.8 and the definition of temperature to determine how the thermal energy U of an ideal monatomic gas depends on its temperature T.

Translation   Let N be the number of gas molecules and V the gas's volume.

Model   We will assume that the gas satisfies the limitations on equation T8.8 and that its volume is fixed (we do not allow it to expand as we increase T). According to chapter T6, an object's temperature T is defined so that $T^{-1} = dS/dU$, where we evaluate the derivative while holding N and V constant.

Solution   We can use $\ln xy = \ln x + \ln y$ to take apart the logarithm in equation T8.8 as follows:

$$S = \frac{3}{2} Nk_B \ln U + \frac{3}{2} Nk_B \ln \left(\frac{8m\,V^{2/3}b}{3h^2 N} \right) - k_B \ln(N!) \tag{T8.9}$$

Note that only the first term depends on U. According to the definition of temperature,

$$\frac{1}{T} = \frac{\partial S}{\partial U} = \frac{\partial}{\partial U} \left(\frac{3}{2} Nk_B \ln U \right) + 0 + 0 = \frac{3}{2} Nk_B \frac{1}{U} \tag{T8.10}$$

since we are supposed to evaluate the derivative while holding V and N

fixed. Multiplying both sides by UT yields

$$U = \frac{3}{2}Nk_BT \qquad \text{(T8.11)}$$

This is consistent with what we found in chapter T2. (Note that the value of b is irrelevant.)

T8.2 Entropy Depends on Volume

Note that an ideal monatomic gas's multiplicity and entropy both increase as its volume V increases (with U and N fixed). Why? Remember that in our model, we treat each molecule as three independent quantum systems whose states have energies $E_n = \varepsilon n^2$. According to equation T8.2, $\varepsilon \propto 1/L^2 = 1/V^{2/3}$ (since $L^3 = V$). Therefore, as the gas's volume increases, the energies of the states available to a molecule's quantum systems decrease. But if U and N are fixed, the average energy $E_{avg} = U/3N$ per quantum system remains the same, so the number of states available to a quantum system between the energy limits 0 and bE_{avg} *increases*. Thus the number of states available to each molecule increases as L increases.

In fact, since $(V^{2/3})^{3N/2} = V^N$, equation T8.7 tells us that $\Omega \propto V^N$, which implies that the number of states available to each molecule increases in direct proportion to V. We might naively think this is so because the number of spatial positions available to a molecule increases in proportion to the gas's volume. Technically, this is not quite right (the argument in the previous paragraph correctly recognizes that the multiplicity counts *energy* states, not position states), but this simple idea is easy to remember and gives the right answer. Moreover, it correctly predicts the important fact that $\Omega \propto V^N$ for *all* types of ideal gases without all the work of finding a detailed multiplicity function for such gases.

Qualitatively, we know that if we make more volume available to any type of gas, it will spontaneously expand to fill the empty space. This fact alone argues that any gas's entropy increases as its volume increases.

Why a gas's entropy increases when it expands

$\Omega \propto V^N$ for all types of gases

Example T8.2

Problem Figure T8.2 shows N gas molecules constrained by a barrier to be in the left half of a box (the right side contains a vacuum). When we remove the barrier, the gas spontaneously expands to fill the box. How much does the gas's entropy increase in this process?

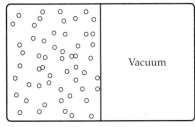
(a) Before barrier is removed

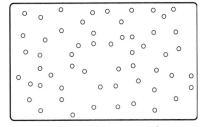
(b) After barrier is removed

Figure T8.2
Free expansion of a gas into a vacuum.

Translation   Let V_i be the gas's initial volume and $V_f = 2V_i$ its final volume.

Model   Nothing happens to the total thermal energy U of the molecules in this expansion; we are simply removing a barrier. We also have a fixed number of molecules N. Therefore nothing changes in equation T8.8 in this process except for V.

Solution   We can use $\ln xy = \ln x + \ln y$ to take apart the logarithm in equation T8.8:

$$S = \frac{3}{2}Nk_B \ln V^{2/3} + \frac{3}{2}Nk_B \ln\left(\frac{8mbU}{3h^2N}\right) - k_B \ln(N!)$$

$$= Nk_B \ln V + \frac{3}{2}Nk_B \ln\left(\frac{8mbU}{3h^2N}\right) - k_B \ln(N!) \qquad \text{(T8.12)}$$

Only the first term here depends on V, which is the only variable changing here. So the change in entropy is

$$\Delta S = Nk_B \ln V_f - Nk_B \ln V_i = Nk_B \ln\left(\frac{V_f}{V_i}\right) = Nk_B \ln 2 \qquad \text{(T8.13)}$$

Evaluation   The units work out, because k_B has the same units as S and both N and $\ln 2$ are unitless. The entropy change is also positive, as we would expect for an irreversible process. Note that, again, the value of b is irrelevant.

T8.3   A General Expression for Entropy Changes

There are situations in which we would like to calculate a macroscopic system's change in entropy in a certain macroscopic process. If we have a nice expression for the system's entropy, this is not a problem. However, exact expressions for S as a function of system properties are not often available, so we need another approach.

The partial derivative $\partial S/\partial U$ is the same as the small change dS in a gas's entropy divided by the small change dU of its internal energy in any process where the gas's volume V and its number of molecules N are constant. Therefore the definition of temperature implies that

$$\frac{1}{T} = \frac{\partial S}{\partial U} \quad \Rightarrow \quad dS = \frac{dU}{T} \quad \text{if } V \text{ and } N \text{ are fixed} \qquad \text{(T8.14)}$$

However, this formula is not correct when V is *not* constant: we have just seen that a gas's entropy increases when it expands, even if U is constant (and thus $dU = 0$). This is so because the definition of temperature *assumes* we are calculating the infinitesimal changes dS and dU while holding all other variables constant, so it does not even apply if V is not constant.

A basic claim about ΔS in adiabatic volume changes

The key to developing a more general version of equation T8.14 is to understand that if a gas changes volume **quasistatically** (slowly enough that the gas is almost in equilibrium at every point during the volume change) and adiabatically (so that no heat flows into or out of the gas), then its change in entropy during the process turns out to be *zero*:

$$\Delta S = 0 \quad \text{for an quasistatic, adiabatic volume change} \qquad \text{(T8.15)}$$

We can see that this is true for a monatomic gas as follows. In section T3.6, we saw that in a sufficiently slow adiabatic process, $TV^{2/3}$ is constant. Since

$U \propto T$ for any ideal gas (assuming N is fixed), this means that $UV^{2/3}$ is constant. But equation T8.8 implies that if $UV^{2/3}$ and N are constant, then S is constant. Although the mathematics is more complicated, the same statement is true for diatomic and polyatomic gases as well.

In an adiabatic volume change, a gas's thermal energy U changes due to work energy flowing into the gas (during a compression) or out of the gas (during an expansion). An object's entropy S generally depends on its thermal energy U. But in an *adiabatic* volume change, the entropy change due to the gas gaining or losing energy is exactly canceled by the change in entropy due to the volume change. For example, in an *expansion* we have seen that the energies of quantum states available to the gas molecules decrease. If U were fixed, there would then be more ways to distribute the gas's energy among its molecules, and thus the gas's entropy would increase. But in an *adiabatic* expansion U decreases (because the gas does work on its surroundings as it expands) exactly in proportion with the decrease in state energies, so the number of ways to distribute that energy (and thus the entropy) does not change.

Energy changes due to quasistatic work do not count

We see that in this case, then, *any change in U due to quasistatic work has no effect on the gas's entropy*. This provides the foundation that enables us to calculate entropy changes for gas processes involving volume changes. We will assume that we can treat any such process as an adiabatic volume change plus an additional transfer of thermal energy in the form of heat Q or other kinds of work. Since we know that the adiabatic part of the process does not change the gas's entropy, it follows that if the gas's entropy *does* change in a process, then it must be due to thermal energy added to the gas in ways *other* than quasistatic volume changes. Therefore

$$dS = \frac{dU - dW_{qs}}{T} \qquad \text{if } N \text{ is constant and } V \text{ changes quasistatically} \qquad \text{(T8.16)}$$

where dW_{qs} is the work associated with the quasistatic volume change. In many gas processes, this is the *only* kind of work involved. If so, the only other way that energy can be added to the gas is in the form of heat, so

$$dS = \frac{dQ}{T} \qquad \text{(T8.17)}$$

A basic formula to use to compute entropy changes

Purpose: This equation tells us how to calculate a system's entropy change dS during an infinitesimal process.

Symbols: dQ is the infinitesimal heat added to the system, and T is the system's temperature.

Limitations: dQ must be small enough that T does not change significantly during the process. The number of molecules N in the system must be constant, its volume V must change quasistatically, and there must be no work involved *other* than that due to the volume change.

We will spend the remainder of the chapter exploring the uses of equation T8.17.

We derived equation T8.17 while assuming that our object was an ideal gas. In practice, however, we can use equation T8.17 for solids and liquids as well. Since most solids and liquids change volume only slightly, if at all, during typical processes, we can treat V as fixed. If N is also fixed, then equation T8.16 applies. However, as long as no *other* kind of work energy enters the solid or liquid in question, then $dU = dQ$, meaning that equations T8.16 and T8.17 are equivalent.

Please note the limitations on this equation

You should carefully note equation T8.17's limitations. When a process involves energy transfers to thermal energy that are not heat and not due to quasistatic volume changes (e.g., when gasoline explodes in an automobile engine cylinder), equation T8.17 does not apply. Equation T8.17 also does not cover sudden or violent volume changes. Surprisingly, it turns out that with a certain amount of cleverness, we can get around these limitations and use equation T8.17 indirectly in such processes, even though it does not strictly apply. We'll see how in section T8.6.

Exercise T8X.3

Which of the following processes are consistent with the limitations on equation T8.18, and which are not?

a. A stone is thrown into a pond.
b. Soup is slowly heated on a stove.
c. A cup of coffee is gently stirred.
d. A bubble slowly rises in a lake.
e. A gas is slowly compressed in a cylinder while its temperature is held fixed.

T8.4   Constant-Temperature Processes

Equation T8.17 strictly applies only to heat transfers that are "sufficiently small" that the temperature does not change significantly during the process. In general, as we transfer heat to an object, its temperature will change. We cannot calculate ΔS for a finite heat transfer ΔQ unless we take account of the changing temperature by doing an integral:

$$dS = \frac{dQ}{T} \quad \Rightarrow \quad \Delta S = \int_{\text{process}} \frac{dQ}{T} \quad \begin{array}{l}\text{quasistatic volume} \\ \text{change, no other} \\ \text{work done, } N \text{ fixed}\end{array} \quad \text{(T8.18)}$$

We will discuss the procedure for evaluating such integrals in section T8.5.

If, on the other hand, the temperature of an object remains approximately constant during the process, then whatever amount of heat is transferred during the process *is* "sufficiently small" and we can apply equation T8.17 directly:

The entropy change when temperature is approximately constant

$$\Delta S = \frac{Q}{T} \quad \text{if } T \approx \text{constant during a process} \quad \text{(T8.19)}$$

Under what circumstances will T be approximately constant? At the end of chapter T1, we discussed how the change dU in an object's thermal energy as its temperature changes by dT is given by

$$dU = mc\, dT \quad \text{(T8.20)}$$

where m is the object's mass and c is its specific heat (note that specific heats for various substances are listed on the inside front cover of this text). Imagine an object so massive that it can absorb or supply a significant amount of heat dU while suffering only the tiniest change in temperature dT. The technical term for such an object in thermal physics (as we saw in chapter T6) is a **reservoir.** Thus, we can use T8.19 to

Three practical situations where $T \approx$ constant

1. Compute the entropy change of a *reservoir* absorbing or supplying heat.
2. Compute the entropy change of something in thermal contact with a reservoir, and thus whose temperature is the same as that of the reservoir.

In both cases, the presence of the reservoir ensures that the temperature T appearing in equation T8.19 does not change significantly during the process.

There is a third (unrelated) case in which we can use equation T8.19. We saw in unit C that the thermal energy we have to add or remove from a substance while it undergoes a **phase change** (e.g., from a solid to a liquid) is

$$\Delta U = \begin{cases} -mL & \text{(gas to liquid, liquid to solid)} \\ +mL & \text{(solid to liquid, liquid to gas)} \end{cases} \qquad \text{(T8.21)}$$

where m is the mass of the substance and L is its **latent heat** (latent heats for various substances are also listed on the inside front cover). Since the temperature of the substance remains *constant* until the phase change is complete, we can use equation T8.20 to

3. Compute the entropy change of a substance during a phase change.

These three cases, then, are the most common practical situations in which we can apply equation T8.20. The following examples illustrate its use.

Example T8.3

Problem   Imagine that we have a bathtub full of water at 20°C and we place a 1.0-kg stone in it whose original temperature is 95°C. What is the entropy change of the water in this case?

Model   The tub of water probably contains hundreds of kilograms of water, and its temperature change in this process will be tiny compared to that of the stone. This means that the final equilibrium temperature of both will be about 20°C.

Translation   According to the inside front cover, the specific heat of granite is about 760 J kg^{-1} K^{-1}. Let Q_w and Q_s be the heat flowing into the water and stone respectively, and let T_w and T_s be the temperatures of the water and stone, respectively.

Solution   The heat energy Q_s flowing into the stone is

$$\Delta Q_s = m_s c_s \, \Delta T_s = (1.0 \text{ kg})(760 \text{ J kg}^{-1}\text{K}^{-1})(-75 \text{ K}) = -57{,}000 \text{ J} \quad \text{(T8.22)}$$

This is negative, so heat is actually flowing *out* of the stone. The water gains this energy, so the water's change in entropy is

$$\Delta S = \frac{\Delta Q_w}{T_w} = \frac{+|\Delta Q_s|}{T_w} = \frac{+57{,}000 \text{ J}}{293 \text{ K}} = +195 \text{ J/K} \qquad \text{(T8.23)}$$

Evaluation   This is positive, as we would expect for something gaining thermal energy.

Example T8.4

Problem   Imagine that we have a cylinder containing an ideal gas in good thermal contact with a reservoir at 32°C. Imagine that we slowly compress the gas, doing 45 J of work on it while its temperature remains constant. What is the change in entropy of the gas? What is the change in entropy of the reservoir?

Translation   Let Q_g and Q_R be the heat flowing into the gas and reservoir, respectively, and let T_g and T_R be the temperatures of the gas and reservoir, respectively.

Model   Since the gas's temperature does not change, its thermal energy U does not change. Therefore any work energy that it gains in this compression must flow out to the reservoir in the form of heat.

Solution   The gas therefore *loses* 45 J of heat in this process, so its entropy change [noting that $T_g = (273 + 32)\ \text{K} = 305\ \text{K}$] is

$$\Delta S_g = \frac{Q_g}{T_g} = \frac{-45\ \text{J}}{305\ \text{K}} = -0.15\ \text{J/K} \tag{T8.24}$$

(Remember that the work energy it gains in the "slow" compression doesn't count!) The 45 J that the gas loses flows *into* the reservoir, so its entropy change is

$$\Delta S_R = \frac{Q_R}{T_R} = \frac{+45\ \text{J}}{305\ \text{K}} = +0.15\ \text{J/K} \tag{T8.25}$$

Evaluation   Note that the *net* entropy change of the system in this process is zero.

Example T8.5

Problem   Imagine that 120 g of ice at 0°C melts to a puddle of water at 0°C on a surface in a room where the temperature is 28°C (see figure T8.3). What is the change in the entropy of the water?

Translation   Let Q_w and T_w be the heat flowing into the ice and its temperature, respectively. Let m be the mass of the ice and L its latent heat.

Model   Note that the temperature of the ice does not change as it melts to water, so we can use equation T8.19.

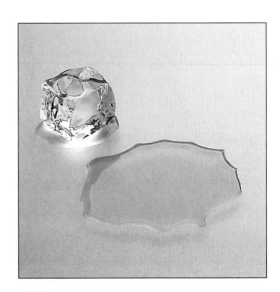

Figure T8.3
What is the entropy change?

Solution   According to the table on the inside front cover, the latent heat associated with the transformation of solid to liquid water is 333 kJ/kg, so the total heat that the ice must absorb from the warm room is

$$Q_w = dU = +mL = (0.12 \text{ kg})(333 \text{ kJ/kg}) = 40 \text{ kJ} \qquad \text{(T8.26)}$$

Its entropy change is therefore

$$\Delta S_w = \frac{Q_w}{T_w} = \frac{+40,000 \text{ J}}{273 \text{ K}} = +150 \text{ J/K} \qquad \text{(T8.27)}$$

Note that the temperature of the room is irrelevant here.

Evaluation   Note that the entropy change is positive, as we should expect when a system's internal energy increases.

Exercise T8X.4

If the table surface in example T8.5 acts as a reservoir, find its entropy change. Does the *total* entropy of water and table increase?

Exercise T8X.5

Note that in each of examples T8.3 through T8.5, I carefully converted temperatures to *kelvins* before computing the entropy. Further, note that if I had divided by temperatures in degrees Celsius, I would have gotten *very* different answers. Why must we use Kelvin temperatures when we use equation T8.19?

T8.5   Handling Changing Temperatures

Now let us consider cases in which we would like to compute the entropy change of an object as it undergoes a heat transfer that changes its temperature significantly. Assume the object has specific heat c and mass m. If the object's volume does not change (much) during the process, then no significant work is done and

$$dQ \approx dU = mc\,dT \qquad \text{if } no \text{ work is done} \qquad \text{(T8.28)}$$

Plugging this into equation T8.19, we get

$$\Delta S = \int_{\text{process}} \frac{dQ}{T} = \int_{\text{process}} \frac{mc\,dT}{T} \qquad \text{if no work is done} \quad \text{(T8.29)}$$

Note that if we consider c to be a function of temperature, this is simply an integral over some function of the object's changing temperature.

If the object's specific heat is approximately independent of temperature (as it often is over reasonably small temperature ranges), then we can pull both m and c out in front of the integral. Integrating the dT/T that remains from the object's initial temperature T_i to its final temperature T_f, we get

$$\Delta S = mc \ln\left(\frac{T_f}{T_i}\right) \qquad \text{if no work is done and } c \approx \text{constant} \qquad \text{(T8.30)}$$

The change in entropy for a heat exchange with no work involved

Exercise T8X.6

Verify equation T8.30.

Example T8.6

Problem   Imagine that I place an aluminum block with mass $m_b = 260$ g and an initial temperature $T_b = 89°C$ into a cup of water with mass $m_w = 320$ g and an initial temperature $T_w = 22°C$, and I allow the two to come to thermal equilibrium. What is the entropy change of the water in this process?

Translation   Let c_b and c_w be the specific heats of the water and aluminum block, respectively.

Model   We will assume that c_b and c_w are constant over the temperature ranges involved, so that we can use equation T8.30.

Translation and Solution   The first step is to find the final equilibrium temperature T_f. The heat flowing out of the metal goes into the water; so if Q_b and Q_w represent the heat flowing into the block and water, respectively, then $Q_b = -Q_w$. This means that

$$m_b c_b (T_b - T_f) = Q_b = -Q_w = m_w c_w (T_w - T_f) \qquad (T8.31)$$

Solving for T_f, we get

$$(m_w c_w + m_b c_b) T_f = m_w c_w T_w + m_b c_b T_b \quad \Rightarrow \quad T_f = \frac{m_w c_w T_w + m_b c_b T_b}{m_w c_w + m_b c_b}$$

$$(T8.32)$$

Using $c_w = 4186$ J kg^{-1} K^{-1} and $c_b = 900$ J kg^{-1} K^{-1} (see inside front cover) and converting $T_w = 22°C$ to $(273 + 22)$ K $= 295$ K (similarly, $T_b = 362$ K), we get

$$T_f = \frac{(0.32 \text{ kg})(4186 \text{ J kg}^{-1}\text{K}^{-1})(295 \text{ K}) + (0.26 \text{ kg})(900 \text{ J kg}^{-1}\text{K}^{-1})(362 \text{ K})}{(0.32 \text{ kg})(4186 \text{ J kg}^{-1}\text{K}^{-1}) + (0.26 \text{ kg})(900 \text{ J kg}^{-1}\text{K}^{-1})}$$

$$= 305 \text{ K} \qquad (T8.33)$$

(See problem T1S.7 for another way to compute the equilibrium temperature.)

Model (part II)   Since $T_f = 305$ K $= 42°C$ is significantly different from the water's initial temperature of 22°C, we need equation T8.30 to compute the water's entropy change.

Solution (part II)   The change in the water's entropy is thus

$$\Delta S_w = m_w c_w \ln\left(\frac{T_f}{T_w}\right) = (0.32 \text{ kg})(4186 \text{ J kg}^{-1} \text{ K}^{-1}) \ln\left(\frac{305 \text{ K}}{295 \text{ K}}\right) = +45 \text{ J/K}$$

$$(T8.34)$$

Evaluation   The units are correct, and the water's entropy change is positive. This is plausible, because the water is gaining energy here.

Exercise T8X.7

Did we have to use absolute temperatures in equation T8.33 (that is, would using temperatures in degrees Celsius make any difference in the final result)? Did we have to use absolute temperatures in equation T8.34?

Exercise T8X.8

Show that the block's entropy change is $-40\,\text{J/K}$. (This means that the net entropy change for the block/water system is $+5\,\text{J/K}$: An *increase* is to be expected in a spontaneous heat transfer process.)

T8.6   Non-quasistatic Processes

As I've said before, equations T8.17 and its descendants T8.19 and T8.30 do not apply to processes in which volume changes are not quasistatic or if work other than that resulting from quasistatic volume changes is done during the process. In many cases of such processes, however, we can actually use these equations, employing a clever argument to do a kind of "end run" around the process.

Our clever argument hinges on the fact that the entropy of any system is defined in terms of the current *macrostate* of that system. Therefore any change in entropy experienced by the system as it goes from one macrostate to another *will depend on its initial and final macrostates alone,* and not at all on the process by which we got from one to the other!

An object's entropy depends on its macrostate alone

Consider a system that goes from one macrostate to another via a process that is *not* quasistatic, or involves work *not* due to a quasistatic expansion or compression, or both. Say that we would like to compute the entropy change of the system in this process. If we can imagine *any* hypothetical process satisfying the restrictions on equation T8.17 that takes us from the same initial macrostate to the same final macrostate, we can calculate the entropy change using this hypothetical process (which we will call a **replacement process**) instead of the actual process: we *must* get the same answer that we would have gotten if we could have done the calculation for the actual process (see figure T8.4).

Using a *replacement process* to calculate an entropy change

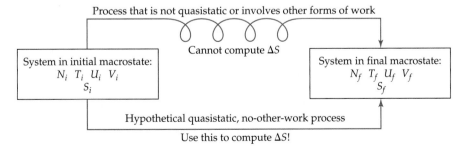

Figure T8.4

How to do an end run around a process violating the restrictions on equation T8.18: invent a process that goes from the same initial to the same final macrostate, and use that process to compute ΔS. Since ΔS only depends on the initial and final macrostates, it does not matter what process you use to compute it.

Example T8.7

Problem   Imagine that I have a basin containing 3 kg of water at 22°C, and I pour a pitcher containing 2 kg of water at 22°C into it from a height of 0.3 m (see figure T8.5). What is the entropy change of the system of both containers of water?

Figure T8.5
What is the entropy change?

Translation and Model   When the 5 kg of water falls into the basin, its gravitational potential energy is converted to kinetic energy and then to thermal energy in the basin. This thermal energy gain is *not* due to a quasistatic volume change, nor is it heat entering the system (there are no temperature differences involved). Therefore, equation T8.17 does not apply.

The initial macrostate of our system is 5 kg of water (3 kg in the basin and 2 kg in the pitcher) at $T = 22$°C and thermal energy U_i. Our final state is the same of water at a slightly higher temperature and a final thermal energy of

$$U_f = U_i + mgh = U_i + (2 \text{ kg})(9.8 \text{ m/s}^2)(0.3 \text{ m}) = U_i + 5.9 \text{ J} \quad (T8.35)$$

Imagine gently lowering the pitcher to the level of the basin and mixing the water in slowly and gently (avoiding adding *any* work as much as possible) and then adding $Q = 5.9$ J of heat to the water. This replacement process will bring the water to the *same* final state without involving any non-quasistatic work.

Since it would take $(5 \text{ kg})(4186 \text{ J} \cdot \text{kg}^{-1}\text{K}^{-1})(1 \text{ K}) > 20{,}000$ J to raise the temperature of the water by even 1 K, our 5.9 J will not change the water's temperature significantly. Therefore, we can use equation T8.19 to compute the water's entropy change during the replacement process.

Solution   Therefore

$$\Delta S = \frac{Q}{T} = \frac{+5.9 \text{ J}}{(273 + 22) \text{ K}} = +0.02 \text{ J/K} \qquad \text{(T8.36)}$$

The water's entropy change during the actual process *must* be the same.

Example T8.8

Problem   Use a replacement process to calculate the entropy change for the sudden expansion process discussed in example T8.2, and show that you get the same answer.

Model and Translation   The volume change here is not even remotely quasistatic, so equation T8.17 does not directly apply. However, since the gas expands into a vacuum, it does no work and there is no time for heat to flow, so the gas's thermal energy U is unchanged in this process. The initial and final macrostates of the gas are thus specified by U, V_i, N and U, V_f, N, respectively, where V_i and $V_f = 2V_i$ are the initial and final volumes, N is the fixed number of molecules involved, and U is the gas's fixed thermal energy. One replacement process that takes us from the same initial macrostate to the same final macrostate is the following. Imagine that we use a piston to gradually allow the gas to expand to its final volume. As the gas expands against the piston, though, work energy will flow out of the gas, leading to a decrease in its thermal energy U. We want the final energy of the gas to be the same as it was originally, so we need to add heat to replace the work energy lost. We can do this in a particularly easy way by putting the gas in thermal contact with a reservoir at 22°C. Since the thermal energy U of an ideal gas depends on T and N but not V, keeping the gas's temperature fixed will automatically add whatever heat is needed to keep its internal energy U fixed. If we calculate the entropy change for this replacement process, it should be the same as the entropy change for the original process.

Solution   The heat added in this replacement process will be equal to the work that the gas does as it expands isothermally. According to equation T3.12, the heat needed is

$$Q = -W = +Nk_B T \ln\left(\frac{V_f}{V_i}\right) \qquad \text{(T8.37)}$$

Since the temperature of the gas is a constant 22°C = 295 K in our replacement process, we can now use equation T8.19 to calculate the gas's change in entropy:

$$\Delta S = \frac{Q}{T} = +\frac{Nk_B T}{T} \ln\left(\frac{V_f}{V_i}\right) = Nk_B \ln 2 \qquad \text{(T8.38)}$$

Evaluation   This is the same as we found in example T8.2.

Exercise T8X.9

Compute a numerical value for ΔS in example T8.8 for 1 mol of gas undergoing such an expansion.

TWO-MINUTE PROBLEMS

T8T.1 Consider a container holding 1 million helium molecules at 300 K. If we double the gas's temperature to 600 K, by about what factor does the helium's multiplicity increase?
A. $\ln 2$
B. 2
C. $2^{3/2}$
D. $10^6 k_B \ln 2$
E. $2^{1,000,000}$
F. $2^{1,500,000}$
T. Some other factor (specify)

T8T.2 How does the entropy of 1 mol of helium gas in a container at room temperature compare to 1 mol of argon gas in an identical container at the same temperature?
A. The helium gas has greater entropy.
B. The argon gas has greater entropy.
C. Both gases have the same entropy.
D. There is not enough information to answer definitively.

T8T.3 Consider each of the following processes. Is the process consistent (C) or inconsistent (D) with the limitations of equations T8.17 and T8.18 (that N be fixed and that the process involve no work *other* than the work produced by a quasistatic volume change)?
a. A glass of milk is spilled on the floor.
b. A closed vial of gas is heated with a flame.
c. A pan of water boils on the stove.
d. Sunlight is absorbed by a black hat.
e. Water evaporates from your skin.

T8T.4 In a tub 100 kg of water at 30°C absorbs 100 J from an electric heater. We can consider the water to be a reservoir in this process (T or F).

T8T.5 100 g of water is slowly heated from 5 to 25°C. What is its change in entropy? (For water $c = 4186$ J/K.)
A. 582 J/K
B. 29.2 K
C. 139 J/K
D. 6.95 J/K
E. 0
F. Other (specify)

T8T.6 A hot object is brought into contact with a cold object. The hot object's entropy decreases by 1.5 J/K as it comes into equilibrium with the cold object. By how much does the cold object's entropy increase?
A. More than 1.5 J/K
B. Exactly 1.5 J/K
C. Less than 1.5 J/K
D. 0

T8T.7 A sample of helium gas is held at a constant temperature while it is slowly compressed in a cylinder until its volume has decreased by a factor of 4. During this process, the gas's entropy
A. Increases
B. Decreases
C. Remains the same
D. Does something else (specify)

T8T.8 A stone of mass m is dropped from a height h into a bucket of water. What would be a suitable replacement process for this non-quasistatic process?
A. Lower the stone gently into the water.
B. Lower the stone gently into the water, and then heat the water until it has mgh more energy than it had before.
C. Lower the stone gently into the water, and then stir the water gently to give it mgh more energy.
D. Raise the water up to height h, gently put the stone in it, and then lower both back to the ground.
E. Either (B) or (C).
F. None of the above.

T8T.9 A sample of helium gas with an initial temperature of T_i in an insulated cylinder with initial volume V_i is suddenly and violently compressed to one-half its volume. The absolute temperature of the gas after the compression is observed to have doubled. (Note: In a *quasistatic* adiabatic compression of a monatomic gas, $TV^{2/3} = $ const.) A suitable replacement process for this process would be
A. A slow isothermal compression from V_i to $\frac{1}{2}V_i$.
B. A slow adiabatic compression from V_i to $\frac{1}{2}V_i$.
C. Heating the gas to double its temperature followed by a slow adiabatic compression from V_i to $\frac{1}{2}V_i$.
D. A slow adiabatic compression from V_i to $\frac{1}{2}V_i$ followed by heating to bring the temperature to $2T_i$.
E. A slow adiabatic compression from V_i to $\frac{1}{2}V_i$ followed by cooling to bring the temperature to $2T_i$.
F. None of the above.

HOMEWORK PROBLEMS

Basic Skills

T8B.1 Starting with equation T8.8, find an expression for the entropy of an ideal monatomic gas as a function of T, V, and N instead of U, V, and N.

T8B.2 Imagine that we compress a sample of N molecules of helium gas to one-half its original volume and simultaneously raise its absolute temperature by a factor of 2. Use equation T8.8 to calculate the change in the gas's entropy in this process as a multiple of Nk_B. (*Hint*: $U \propto T$.)

T8B.3 Imagine that we add 12 J of heat to 10 kg of water at 15°C. Argue that this amount of energy will not change the temperature of the water very much. Compute the entropy change of the water.

T8B.4 Imagine that we put a small ice cube in a bucket containing 5 kg of water at 22°C. As the ice melts, it absorbs 35 J of energy from the water. By about how much has the entropy of the water decreased?

T8B.5 Imagine that the temperature of a 220-g block of aluminum sitting in the sun increases from 18°C to 26°C. By about how much has its entropy increased?

T8B.6 Imagine that a 330-g cup of water is heated in a hot pot from 12°C to 92°C. By about how much has its entropy increased?

T8B.7 In a pan on a stove 125 g of water at 100°C is converted to steam. What is the entropy change of the water that is now steam? (The latent heat of boiling water is 2256 kJ/kg.)

T8B.8 A puddle containing 0.80 kg of water at 0°C freezes on a cold night, becoming ice at 0°C. What is the entropy change of the water that is now ice? (The latent heat of freezing water is 333 J/K.)

Synthetic

T8S.1 *Stirling's approximation* states that $\ln(N!) \approx N \ln N - N$. Plug this and the correct value of b into equation T8.8 and do some algebra to show that the entropy of an ideal monatomic gas can also be written as

$$S(U, V, N) = Nk_B \ln\left[\frac{V}{N}\left(\frac{4\pi mU}{3Nh^2}\right)^{3/2}\right] + \frac{5}{2}Nk_B$$

(T8.39)

This equation is called the *Sackur-Tetrode equation*; it is more convenient than equation T8.8 if one is interested in how S varies with N.

T8S.2 Argue that $n = (8mbU/3NL^2h^2)^{1/2} \gg 1$ at room temperature for any monatomic gas for essentially any reasonable value of L, irrespective of the value of b (as long as b is of the order of magnitude of 1). (*Hints*: How does U depend on T for a monatomic gas? What monatomic gas will make n smallest? Factor out b and L, and compute the rest; then think about how small L would have to be to make n small.)

T8S.3 In newtonian mechanics, we can describe a particle's *state* by specifying its position vector and momentum

vector. We can make a newtonian estimate of a monatomic gas's multiplicity as follows. A molecule's kinetic energy is given by $K = p^2/2m = (p_x^2 + p_y^2 + p_z^2)/2m$. On average, therefore, we would expect $p_x^2/2m = p_y^2/2m = p_z^2/2m = \frac{1}{3}K_{avg}$. Let us *assume* that the number of "x-momentum states" available to a molecule is *proportional* to the value of $|p_x|$ when $p_x^2/2m = \frac{1}{3}K_{avg}$, and the same is true for the number of y momentum and z momentum states. Also, assume that the number of "position states" available to a molecule is proportional to the gas's volume V.

(a) Argue that these assumptions imply that the number of position and momentum states available to a molecule is therefore

$$\Omega_{mol} = \frac{V}{H^3}\left(\frac{2}{3}mK_{avg}\right)^{3/2}$$

(T8.40)

where H is some unknown constant of proportionality with units of momentum × length, or kg m²/s.

(b) Argue therefore that the multiplicity of the entire gas can be written as

$$\Omega = \left(\frac{2mV^{2/3}U}{3NH^2}\right)^{3N/2}$$

(T8.41)

(c) Explain why if we had used this result instead of equation T8.8 in examples T8.1 and T8.2 and in section T8.3, we would have come to exactly the same conclusions, even knowing nothing about the value of H.

(d) Note that there is no good justification in newtonian mechanics for the $1/(N!)$ factor in equation T8.8. Newtonian particles are *always* distinguishable in a certain sense, because in principle we can keep track of their trajectories (something we cannot do in quantum mechanics). But ignoring the $1/(N!)$ in equation T8.8, what value would H here have to have to agree with equation T8.8? (Express your answer in terms of h.)

T8S.4 Imagine that you use an electrical wire to add 28 J of energy to a basin containing 65 kg of water at 18°C. What is the approximate entropy change of the water? (*Hint*: Argue that the water is essentially a reservoir.)

T8S.5 Imagine that you put a 1.0-kg block of aluminum whose initial temperature is 80°C into the ocean at a temperature of 5°C. The ocean and block come into thermal equilibrium. What was the entropy change of the block in this process? That of the ocean?

T8S.6 Imagine that 22 g of helium gas in a cylinder expands quasistatically while in contact with a reservoir at a temperature of 25°C and does 85 J of work on its surroundings in the process. What is the entropy change of the gas? Of the reservoir?

T8S.7 Imagine that you allow 0.40 mol of nitrogen to expand from a volume of 0.005 m³ to a volume of 0.015 m³ while the gas is held at a constant temperature of 304 K. By how much does the entropy of the gas change?

T8S.8 Imagine that you put a block of copper with a mass of 320 g and an initial temperature of −35°C into an insulated cup containing 420 g of water at 22°C. After everything has come to equilibrium, by how much has the entropy of both the water and the copper changed? Explain why the coating of ice that initially forms around the copper is irrelevant.

T8S.9 Imagine that you have a gas confined in an insulated cylinder by a piston. Imagine that you suddenly push hard on the piston, compressing the gas to one-half its volume. Then you slowly let the gas expand back to its original volume. Will the gas have the same temperature as when you started? Will its entropy be the same as before? Would your answers be different if you had slowly compressed the gas and then slowly expanded it? Explain your responses carefully. (*Hint:* When you press the piston in suddenly, do you think that you exert more, less, or the same force on it that you would if you were to compress the gas slowly?)

T8S.10 A 3-g bullet flying at 420 m/s hits a 2.2-kg aluminum block and embeds itself in the aluminum. After everything has come to equilibrium, how much has the entropy of the aluminum block increased? Be sure to describe the replacement process that you use to actually calculate the entropy change.

Rich-Context

T8R.1 Repeat the argument in section T8.1 for the case of the Einstein solid, and show that the multiplicity of the Einstein solid is indeed

$$\Omega(U, N) = \left(\frac{bU}{3N\varepsilon}\right)^{3N} \tag{T8.42}$$

as claimed in section T8.1. Be very careful in your solution to pinpoint the places where the two derivations diverge. In particular, why is the exponent 3N in this case, but 3N/2 in the gas case? Why does the volume appear in the gas case, but not in the Einstein solid case? Why is there no factor of 1/(N!) in the Einstein solid case? (*Hint:* The molecules in a solid

are not free to roam around, so we can uniquely identify each molecule by its lattice position in the solid.)

T8R.2 A high diver dives from the top of a 35-m tower into a pool of water. After the splashing settles down (but before any heat has a chance to be transferred from the water to the diver or vice versa), what is the entropy change of the water in the pool? Explain why one cannot calculate the entropy change for this process directly, and describe the replacement process that you used to calculate the entropy change. Also describe any approximations or estimations that you have to make.

Advanced

T8A.1 Consider the following paradox, known as *Gibb's paradox*. Imagine a box of volume 2V divided in half by a barrier into two chambers with volume V. We fill each chamber with the same number N of molecules of helium. If we remove the barrier, the gas originally in each chamber now has twice the volume available to it, so it seems plausible that the entropy of the combined system would increase. However, it is also clear that we can simply reinsert the barrier to restore the original situation. This would be impossible if the entropy of the combined system had *really* increased when the barrier was removed, as then reinserting the barrier would cause the entropy of the system to decrease, in violation of the second law. So does the entropy of the gas increase when we remove the barrier or not?

(a) Use equation T8.8 to show that the entropy of the combined system *would* increase if we were to omit the $k_B \ln(N!)$ term.

(b) Use the Sackur-Tetrode equation (see problem T8S.1) to show that the entropy of the combined system does *not* increase when we include the $k_B \ln(N!)$ term. [Note that the Sackur-Tetrode equation is equivalent to equation T8.8 in the large-N limit: It simply absorbs the $k_B \ln(N!)$ term into the expression in a more convenient form.]

(c) So we *must* have the $k_B \ln(N!)$ term in the entropy [and thus the 1/(N!) term in the multiplicity] to resolve this paradox. (This was in fact an argument that physicists gave for including this term before the invention of quantum mechanics.) Explain *qualitatively* (using physical ideas and considering the importance of this term) why removing the barrier does not change the system's entropy in spite of the volume change.

ANSWERS TO EXERCISES

T8X.1 If $n \gg 1$ (but *only* if $n \gg 1$), we can ignore the difference between the value $(bU/3N\varepsilon)^{1/2}$ and the nearest integer.

T8X.2 Since $V = L^3$, $L^2 = V^{2/3}$. Plugging this into the definition of ε in equation T8.2, we get

$$\varepsilon = \frac{h^2}{8mL^2} = \frac{h^2}{8mV^{2/3}} \tag{T8.43}$$

Plugging this into equation T8.5 yields equation T8.7. Since $\ln(x/y) = \ln x - \ln y$ and $\ln x^z = z \ln x$, we have

$$S \equiv k_B \ln \Omega \approx k_B \ln \left[\frac{1}{N!} \left(\frac{bU}{3N\varepsilon} \right)^{3N/2} \right]$$

$$= k_B \ln \left[\left(\frac{bU}{3N\varepsilon} \right)^{3N/2} \right] - k_B \ln (N!)$$

$$= \frac{3}{2} N k_B \ln \left(\frac{bU}{3N\varepsilon} \right) - k_B \ln (N!)$$

$$= \frac{3}{2} N k_B \ln \left(\frac{8m V^{2/3} U}{3Nh^2} \right) - k_B \ln (N!) \qquad \text{(T8.44)}$$

using equation T8.43 for ε.

T8X.3 Consistent: (b), (d), (e); inconsistent: (a), (c).

T8X.4 Let the heat flowing into the table be Q_t and let the table's fixed temperature be $T_t = (273 + 28)\,\text{K} = 301\,\text{K}$. Since the energy gained by the ice comes from the table, $Q_t = -Q_w = -40\,\text{kJ}$. The table's entropy change is therefore

$$\Delta S = \frac{Q_t}{T_t} = \frac{-40,000\,\text{J}}{301\,\text{K}} = -133\,\text{J/K} \qquad \text{(T8.45)}$$

The total entropy of the system increases by about 17 J/K. (We would expect an entropy increase, since the process is irreversible.)

T8X.5 Equation T8.17 is ultimately based on the definition of temperature (equation T8.14). In chapter T6, we saw that this definition coincides with the absolute temperature scale defined by the ideal gas thermoscope. So T in any equation based on equation T8.14 has to be *absolute* temperature.

T8X.6 If m and c are constant, then

$$\int_{T_i}^{T_f} \frac{mc}{T}\, dT = mc \int_{T_i}^{T_f} \frac{dT}{T} = mc(\ln T_f - \ln T_i)$$

$$= mc \ln \left(\frac{T_f}{T_i} \right) \qquad \text{(T8.46)}$$

T8X.7 Since equation T8.33 is based on a rearrangement of equation T8.31, where only temperature *differences* appear, it does not matter whether we use temperatures in degrees Celsius or kelvins (1 K = 1°C of temperature difference). In equation T8.34, on the other hand, we must use absolute temperatures. (I generally try to use absolute temperatures exclusively to reduce my chances of error.)

T8X.8 The block's change in entropy is

$$\Delta S_b = mc \ln \left(\frac{T_f}{T_b} \right)$$

$$= (0.26\,\text{kg})(900\,\text{J kg}^{-1}\,\text{K}^{-1}) \ln \left(\frac{305\,\text{K}}{362\,\text{K}} \right)$$

$$= -40\,\text{J/K} \qquad \text{(T8.47)}$$

T8X.9 Plugging numbers into equation T8.38, we get

$$\Delta S = N k_B \ln 2 = (6.02 \times 10^{23})(1.38 \times 10^{-23}\,\text{J/K})(\ln 2)$$

$$= 5.8\,\text{J/K} \qquad \text{(T8.48)}$$

T9 Heat Engines

Chapter Overview

Introduction

Heat engines play a number of essential roles in our technological society. In this final chapter of unit T, we explore the limitations that the second law of thermodynamics places on the ability of a heat engine to convert heat energy to useful work.

Section T9.1:   Perfect Engines Are Impossible

Designs for a **perpetual motion machine** that would run endlessly without using fuel fall into two categories: (1) *perpetual motion machines of the first kind,* which violate conservation of energy (the first law of thermodynamics), and (2) *perpetual motion machines of the second kind,* which conserve energy but violate the second law of thermodynamics.

An example of the second category would be a machine that takes heat energy from some thermal reservoir and converts it entirely to useful mechanical energy (work) without doing anything else. During an operating cycle, such an engine would gain entropy along with the heat from the hot reservoir; but since the work it puts out cannot carry away that entropy, the engine's entropy must disappear (in violation of the second law) so that the engine can return to its initial state to begin the next cycle.

Section T9.2:   Real Heat Engines

A real heat engine satisfies the second law by exhausting some of the heat it gets from the hot reservoir to a cold reservoir, as this provides a way to carry away the engine's entropy. One can do this and have energy left over to convert to work if there is a temperature difference between the hot and cold reservoirs.

Section T9.3:   The Efficiency of a Heat Engine

Since some heat must flow to a cold reservoir to carry away an engine's entropy, no engine can convert all the energy it gets from the heat source to useful work. We define an engine's **efficiency** e to be the fraction of the heat from the hot reservoir that the engine converts to useful work (this is the benefit-to-cost ratio). The second law requires that the net entropy change of the system consisting of the hot and cold reservoirs and the engine not be negative for a cycle of the engine. With a little algebra, one can show that this limits a heat engine's efficiency as follows:

$$e \equiv \frac{|W|}{|Q_H|} \leq \frac{T_H - T_C}{T_H} \tag{T9.7}$$

Purpose: This equation describes the limit that the second law of thermodynamics places on the efficiency of a heat engine that taps the heat flow between a hot reservoir at temperature T_H and a cold reservoir at temperature T_C to produce work $|W|$.

Symbols: $|Q_H|$ is the heat extracted from the hot reservoir.

Limitations: Both T_H and T_C must be absolute temperatures. Some heat engines do not have well-defined hot and cold reservoirs; but if you take T_H to be the highest temperature of the engine's working substance during a cycle, the engine's efficiency will not even approach this limit.

Section T9.4: Consequences

This equation has three important consequences: (1) Any temperature difference can be exploited (in principle) to produce work. (2) The maximum efficiency increases as the temperature difference increases. (3) Waste heat is inevitable. These consequences have a variety of important economic and environmental implications, which are discussed in the section.

Section T9.5: Refrigerators

A refrigerator is essentially a reversed heat engine: It *uses* work to pump heat from a cold reservoir at temperature T_C to a hot reservoir at temperature T_H. We define a refrigerator's **coefficient of performance** (COP) to be the ratio of the heat it removes from the cold reservoir to the work required (again, a benefit-to-cost ratio). A calculation similar to that for a heat engine yields the following limit on a refrigerator's performance:

$$\text{COP} \equiv \frac{|Q_C|}{|W|} \leq \frac{T_C}{T_H - T_C} \qquad \text{(T9.12)}$$

Purpose: This equation describes the limit that the second law of thermodynamics imposes on the coefficient of performance of a refrigerator that uses work $|W|$ to extract heat $|Q_C|$ from a cold reservoir at temperature T_C and exhausts energy $|Q_H|$ to a hot reservoir at temperature T_H.

Limitations: Both T_H and T_C must be absolute temperatures. The equation assumes that the refrigerator operates between well-defined reservoirs.

A **heat pump** is a refrigerator that uses the outdoors as a cold reservoir and exhausts its heat into a house. For typical temperature differences, the latter is much larger than the work used, so this is an energy-efficient way to heat a house.

Section T9.6: The Carnot Cycle

The **Carnot cycle** is a sequence of processes using an ideal gas as a working substance that one can use (in principle) to create a maximally efficient heat engine. The sequence is (1) an isothermal expansion with the gas at the same temperature T_H as the hot reservoir, (2) an adiabatic expansion that takes the gas to the temperature T_C of the cold reservoir, (3) an isothermal compression at temperature T_C, and (4) an adiabatic compression that takes the gas temperature back to T_H. The net entropy change of the reservoirs and the gas during this cycle is zero, so an ideal engine based on this cycle would be maximally efficient. One can also create a refrigerator whose COP is maximal by running this cycle in reverse.

However, even a very large Carnot engine operating at anything close to maximal efficiency would generate energy at an impractically small rate. The homework problems discuss engine cycles that are less efficient but more practical.

T9.1   Perfect Engines Are Impossible

Perpetual motion machines

Since the dawn of the Industrial Revolution, inventors have been trying to build a **perpetual motion machine** that would run endlessly without using any fuel. (Obviously, such a machine could make its inventor very rich.) Many attempts to achieve perpetual motion have been made in the past several centuries, some quite ingenious. But all such attempts have failed, and indeed *must* fail. Why?

All designs for perpetual motion machines fall into two broad categories:

1. *Perpetual motion machines of the first kind.* These designs seek to create the energy required for their operation out of nothing, thus violating the law of conservation of energy (the first law of thermodynamics).
2. *Perpetual motion machines of the second kind.* These designs extract the energy required for their operation from sources in a manner that requires the entropy of an isolated system to decrease (thus violating the second law of thermodynamics.)

An example of a perpetual motion machine of the first kind is a self-powered electric car that uses its motion through the air to turn a windmill that recharges the car's batteries, thus allowing it (in principle) to run forever. But since a moving car always dissipates energy to its surroundings (due to various forms of friction) and since no source for that energy is apparent in this design, it must fail.

Perpetual motion machines of the second kind are usually less obviously flawed. For example, the earth's oceans contain an enormous amount of thermal energy. Why not construct a ship engine that soaks up some of this thermal energy (making the ocean a bit cooler) and converts it to mechanical energy, to drive the ship forward? Conservation of energy is not violated here, so it seems more plausible that such an engine might work. Yet such an engine violates the *second* law of thermodynamics. Why?

Flowing work cannot carry entropy

The key to answering this question lies in understanding that *flowing heat carries entropy from one object to another, but flowing work does not.* Consider first heat energy flowing from a hot object to a cold object. The hot object's entropy decreases in this process, but this is fine because, as we have seen, the cold object's entropy increases at least as much in the process. We can therefore think of this process as carrying the hot object's entropy to the cold object and as well as adding a bit more that was created in the process. But if an object does *work* on another object, this does *not* similarly transfer entropy from the first to the second. For example, if the work energy flowing from the first object raises a weight or adiabatically compresses a gas, it does not change the second object's entropy at all.

Now, figure T9.1 shows an abstract diagram of a "perfect" **heat engine** (such as our hypothetical ocean-powered engine). In a given interval of time (say, one cycle of the engine), such an engine would extract heat energy $|Q_H|$ from a thermal reservoir at temperature T_H (the ocean in our case) and turn that energy entirely into mechanical energy, as shown. The schematic diagram illustrates the flow of energy almost as if it were a river of water. This is a useful mental image: since energy is conserved, it does behave much as a flowing indestructible substance.

Why a perfect engine violates the second law

Figure T9.1 helps us to understand why such an engine violates the second law of thermodynamics. Imagine that during part of its cycle, the engine absorbs some heat from the ocean. The ocean's entropy decreases during this process, so the engine's entropy must increase at least as much. If, in the next

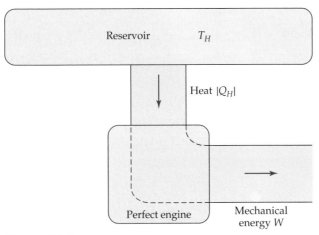

Figure T9.1
A schematic diagram of a perfect heat engine, which extracts a certain amount of heat energy from a reservoir and converts it entirely to useful mechanical energy.

part of the cycle, the engine converts this heat entirely to work, then the energy we got from the ocean flows away, but the entropy remains in the engine. But to repeat the cycle (or maintain its state if it operates in a steady-state mode), the engine must return to its original macrostate and thus to its original entropy. So the engine must get rid of the entropy it got from the ocean somehow. Since we have no energy left to create a heat flow to carry the entropy away, the entropy must spontaneously disappear, in violation of the second law.

T9.2   Real Heat Engines

So how can we construct a heat engine that satisfies the second law? The trick is as follows. Instead of converting *all* the energy from the heat source to work, the engine needs to save some energy to flow as heat from the engine to something colder, and so carry away the entropy the engine got from the heat source. As we will see, as long as the temperature T_C of the cold reservoir where we dump this waste heat is lower than the temperature T_H of the heat source, we will have a bit of energy left over that we can send out of the engine as work. This process is schematically illustrated in figure T9.2.

The presence of a *temperature difference* is crucial here. In the language of the financial analogy of chapter T6, the cold reservoir wants the hot reservoir's money (heat), because its happiness (entropy) will increase more in such a transfer than the hot reservoir's happiness decreases. The cold reservoir wants this transfer so much, in fact, that it is willing to pay a broker (the engine) a small profit to mediate the transfer. The broker's happiness is not changed by this transaction (it is just a job), so if the broker charges too much, the cold reservoir does not get enough money so that its happiness increases more than the hot reservoir's decreases, and the deal is off. The maximum amount that the broker can charge is the amount that exactly balances the happiness increase of the cold reservoir with the happiness decrease of the hot reservoir. If there is no generosity difference (temperature difference) between the reservoirs, the broker cannot charge anything.

A visual representation of what a heat engine does

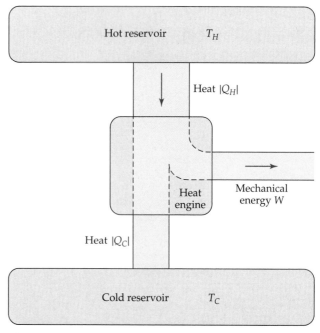

Figure T9.2

A schematic diagram of a real heat engine. A heat engine essentially taps the spontaneous flow of heat from a hot object to a cold object, diverting some of the flow and converting it to mechanical energy.

An example of a heat engine: a steam turbine

A good example of a heat engine is a steam turbine in an electric power plant (see figures T9.3 and T9.4). Heat from some source (nuclear power or burning fossil fuels) converts water in a boiler to steam. (Since heat is constantly supplied to the boiler at a constant temperature by whatever the heat source is, the heat source acts as if it were a constant-temperature hot reservoir.) Water expands greatly when it boils, so this means that the steam is under high pressure. This high-pressure steam is allowed to expand by blowing against the blades of a turbine, which converts some of the internal energy of the steam to mechanical energy (which is ultimately turned into electrical energy by a generator). The spent steam is sent to a condenser, which cools the steam and condenses it back into liquid water by putting it in contact with a cold reservoir. In the old days, a nearby river was used as the cold reservoir, but this often increased the temperature of the water enough to be harmful to life downstream. Most modern power plants employ cooling towers, which essentially use the atmosphere as the cold reservoir. In either case, the recondensed water is then returned to the boiler in essentially its original state, completing the cycle.

Any temperature difference can (in principle) be used to drive a heat engine. During the energy crisis of the 1970s, some research was done on the possibility of generating electrical energy from the temperature difference between the ocean surface and ocean floor (which can be as large as 30 K). The plan for such a power plant would be much like that shown in figure T9.3, except that we would have to use a working substance that boils at a lower temperature than water (ammonia is a possibility). Such a plant would generate electrical power without using any fuel (ultimately, the temperature difference in the ocean is maintained by energy from the sun)! There are, however, some thorny practical and environmental issues associ-

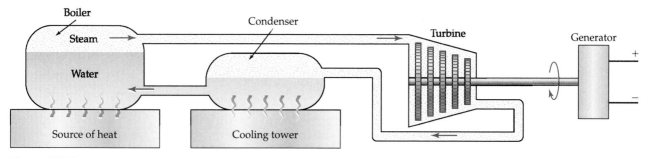

Figure T9.3

A schematic diagram of a steam turbine in an electric power plant. Something (burning fossil fuels, nuclear power) acts as a *hot reservoir*, whose heat boils water, converting it to steam which expands against the blades of the turbine, turning the generator and creating electrical energy. The spent steam is sent to a *condenser* where it gives up heat to the atmosphere (the cold reservoir) and condenses back into water, completing the cycle.

Figure T9.4

The rotating part of the kind of steam turbine used in electric power plants.

ated with doing this on a large scale. (A good discussion of this proposal can be found in Penney and Bharathan, "Power from the Sea," *Scientific American*, January 1987. Figure T9.5 shows a possible schematic diagram of such a plant.)

T9.3   The Efficiency of a Heat Engine

The purpose of any heat engine is to convert to mechanical energy as much of the heat $|Q_H|$ extracted from the hot reservoir (which is typically energy provided by burning precious fuel) as possible. We can quantify how well an engine does this in terms of its **efficiency** e, which is defined to be the ratio of the mechanical energy $|W|$ produced (the benefit) to the heat energy $|Q_H|$ extracted (the cost):

$$e \equiv \frac{\text{benefit}}{\text{cost}} = \frac{|W|}{|Q_H|} \qquad\qquad \text{(T9.1)}$$

Definition of efficiency

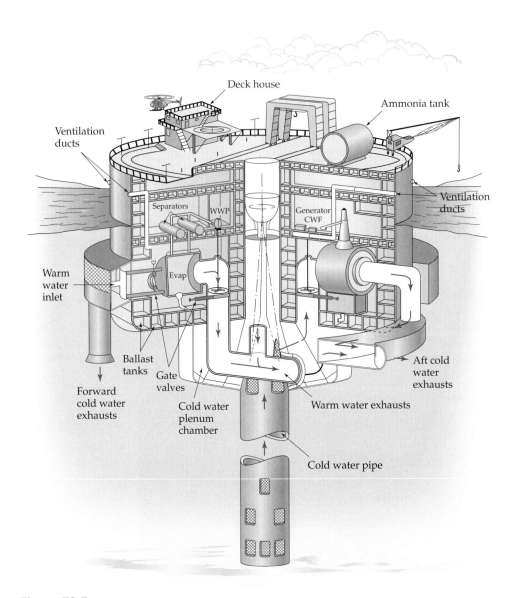

Figure T9.5

An artist's conception of a plant for exploiting oceanic temperature differences to produce electric power. Adapted from R. Ristinen and J. Kraushaar, *Energy and the Environment*, Wiley, New York, 1999, p. 146.

Conservation of energy implies that the work $|W|$ that the engine produces is the difference between the energy $|Q_H|$ it gets from the hot reservoir and the energy $|Q_C|$ it sends to the cold reservoir:

$$|W| = |Q_H| - |Q_C| \tag{T9.2}$$

Exercise T9X.1

Combine equations T9.1 and T9.2 to show that the efficiency can be written

$$e = 1 - \frac{|Q_C|}{|Q_H|} \tag{T9.3}$$

The limit on efficiency imposed by the second law

A perfect engine of the type described in section T9.1 would have an efficiency of $e = 1$: such an engine converts all $|Q_H|$ to W and gives up no heat

$|Q_C|$ to a cold reservoir. As we've seen, a perfect engine violates the second law of thermodynamics, so the efficiency of any real engine will therefore be less than 1.

The question is, how much less? We can in fact calculate this fairly easily. Imagine that one cycle of our heat engine absorbs heat $|Q_H|$ from a hot reservoir at a fixed temperature T_H, produces a certain amount of mechanical energy W, and then gives up heat $|Q_C|$ to a cold reservoir at a fixed temperature T_C. Since both the hot and cold reservoirs are considered to have fixed temperatures, we can calculate the entropy changes of these reservoirs using equation T8.19:

$$\Delta S_H = -\frac{|Q_H|}{T_H} \qquad \Delta S_C = +\frac{|Q_C|}{T_C} \qquad \text{(T9.4)}$$

Note that ΔS_H is *negative* because the hot reservoir loses entropy as it gives up heat. The cold reservoir, on the other hand, *gains* entropy as it absorbs heat. In the formula above, I have used the absolute values of the heat lost or gained to make the signs of ΔS_H and ΔS_C explicit.

Exercise T9X.2

The entropy change of the engine itself during a cycle is zero, and the work created does not carry away any entropy. Argue that requiring that the entropy of the interacting system consisting of the engine and the reservoirs must increase (or at least remain the same) implies that

$$\frac{|Q_C|}{T_C} - \frac{|Q_H|}{T_H} \geq 0 \qquad \text{(T9.5)}$$

Exercise T9X.3

Show that equation T9.5 can be rearranged to read

$$\frac{|Q_C|}{|Q_H|} \geq \frac{T_C}{T_H} \qquad \text{(T9.6)}$$

If you combine equations T9.3 and T9.6, you can show that

$$e \equiv \frac{|W|}{|Q_H|} \leq 1 - \frac{T_C}{T_H} = \frac{T_H - T_C}{T_H} \qquad \text{(T9.7)}$$

Maximum theoretical efficiency of a heat engine

Purpose: This equation describes the limit that the second law of thermodynamics places on the efficiency of a heat engine that taps the heat flow between a hot reservoir at temperature T_H and a cold reservoir at temperature T_C to produce work $|W|$.

Symbols: $|Q_H|$ is the heat extracted from the hot reservoir.

Limitations: Both T_H and T_C must be absolute temperatures. Some heat engines do not have well-defined hot and cold reservoirs; but if you take T_H to be the highest temperature of the engine's working substance during a cycle, the engine's efficiency will not even approach this limit.

Exercise T9X.4

Verify equation T9.7.

This inequality applies no matter how cleverly the engine is designed. Any engine that seeks to do better than this will give up too little heat to the cold reservoir to make that reservoir's entropy gain exceed the hot reservoir's entropy loss, and thus it will violate the second law.

Note that this maximum efficiency is essentially equal to the fraction that the temperature *difference* between the reservoirs is of the temperature of the high-temperature reservoir. As the temperature difference between the reservoirs increases relative to T_H, the maximum possible efficiency goes up. As the temperature of the cold reservoir approaches absolute zero, the efficiency of the engine approaches the perfect value of 1.

The derivation of equation T9.7 described here is one of those derivations that every educated person should know and understand (partly because of its economic and environmental consequences, which we will discuss shortly). Make *sure* that you understand each step of the derivation and that (in the long run at least) you can reproduce the argument if asked.

T9.4   Consequences

Consequences

As I mentioned before, our technological society extensively employs heat engines in a wide variety of applications. Equation T9.7 therefore has several important conceptual, economic, and environmental consequences.

1. *Any temperature difference can be exploited to generate mechanical energy.* Equation T9.7 tells us that if we have two reservoirs at different temperatures $T_H > T_C$, then in principle we can construct a heat engine that exploits that temperature difference to produce mechanical energy. One of the reasons we use fossil fuels at such a tremendous rate is that it is relatively easy to create a substantial temperature difference by burning such fuels. On the other hand, this fact also offers some hope when the fossil fuels eventually run out: *any* means of generating a temperature difference provides an opportunity to convert heat energy to mechanical energy.

2. *The greater the temperature difference, the more efficient the engine.* Equation T9.7 makes clear the advantage in making $T_H - T_C$ as large as possible. Since the most practical low-temperature reservoir available to most engines is the surrounding environment (at $T_C \approx 0°C$ to $25°C$), this usually means making T_H as large as possible. The trade-off is usually that constructing engine parts able to withstand extremely high temperatures is expensive, and there comes a point where it becomes more economical to waste fuel than it is to use very expensive materials to boost efficiency by a few percent.

3. *Energy waste is inevitable.* No heat engine is going to be able to convert all the energy of its fuel to useful mechanical energy: *some* of that energy is necessarily discarded to the cold reservoir. No amount of technological ingenuity will enable one to get around the basic limit on efficiency imposed by the second law of thermodynamics, stated by equation T9.7.

The problem of waste heat

This last issue has some serious environmental consequences. Disposing of the inevitable waste heat generated by large heat engines (or large numbers of small engines) in a way that does not detrimentally affect the environment can be a real challenge. Most cities are substantially warmer than the surrounding countryside partly because of the waste heat being produced within their borders. It has been estimated that the waste heat produced by human activities in the Los Angeles basin now exceeds 7% of the solar energy falling on the basin.

The problem of disposing of waste heat is particularly acute at large electric power plants, where very large heat engines are used to convert heat to electrical power. A large electric plant may need to dispose of several billion joules of waste heat energy into a suitable cold reservoir every second. Directly dumping this kind of energy into a passing river (the most economical method) can have severe environmental consequences. Modern cooling towers operate by taking cold river water, spraying it over the pipes containing the working substance to be cooled, and then allowing the river water to evaporate. Since the latent heat of vaporization of water is fairly large, this carries a lot of heat away for the amount of water vaporized. While this method does not increase the river's temperature, it does reduce the amount of water flowing in the river, which can also have some negative environmental effects (although these are usually less severe than directly dumping the heat in the river).

Exercise T9X.5

In a typical nuclear power plant, heat from the reactor core is used to produce pressurized steam at a temperature of about 300°C (the limit on this temperature is primarily imposed by the dangers of melting the reactor fuel). The temperature of recondensed water leaving the cooling tower is about 40°C. Show that the maximum possible efficiency of the steam turbines in such a plant is about 0.45.

Exercise T9X.6

Due to various unavoidable factors (mechanical friction, the energy needed to pump water around, etc.) the actual efficiency of such a plant is about 0.34. If the plant generates 1000 MW of electrical power (10^9 J/s), show that about 1900 MW of energy has to be given up in the form of waste heat. (This means that nearly two-thirds of the energy in the expensive fuel must be wasted!)

Exercise T9X.7

Imagine that this plant is located on the banks of a major river that is 67 m wide near the plant, that is an average of 3 m deep, and that flows at a rate of about 0.5 m/s. If *all* this water could be routed through the plant and the waste heat could be distributed *evenly* in it, show that the river downstream from the plant will be about 4.5°C warmer than that upstream of the plant. (The specific heat of water is 4186 J kg^{-1} K^{-1}, and the density of water is 1000 kg/m^3.)

Exercise T9X.8

If the water is routed to cooling towers, what fraction of the river has to be evaporated to carry away the waste energy? (The latent energy of vaporization of water is about 2257 kilojoules per kilogram (kJ/kg). You should find a result of less than 1%: this is going to be much less disruptive of the river's ecology than heating it by 5°C.)

Figure T9.6
A hurricane (a powerful natural heat engine).

Since the efficiency of a heat engine improves as the temperature difference involved increases, heat engines that exploit natural temperature differences (e.g., engines that extract power from oceanic temperature differences, geothermal temperature differences, or temperature differences created using solar energy) are often less efficient than fossil-fuel powered engines because natural temperature differences are typically smaller. Since heat energy provided by such natural sources is essentially "free," this does not matter so much, except that the naturally driven engines often need to be bigger and more expensive to produce the same amount of power, and the disposal of waste heat can be a more significant problem (see problem T9S.8 for an example).

This does not mean that you cannot extract a *lot* of energy from such sources. A hurricane is essentially a natural heat engine that gets the motive energy for its winds from relatively small temperature differences between the ocean and the atmosphere (figure T9.6). The efficiency of conversion is small, but the work done by the winds created by such a hurricane can exceed 10^{17} J/day (which is about the energy released by 1000 Hiroshima-sized atom bombs).

T9.5   Refrigerators

What a refrigerator does

A **refrigerator** is a device that would seem at first glance to violate the second law of thermodynamics. We have seen that heat energy naturally and spontaneously flows from hot to cold because the combined entropy of the hot

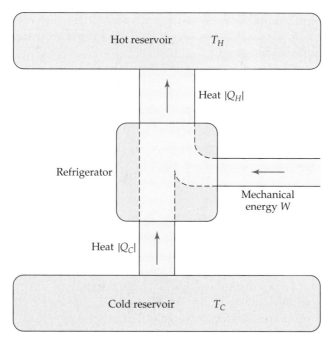

Figure T9.7
A schematic diagram of a refrigerator. Note that a refrigerator is essentially a heat engine operating in reverse: Instead of producing mechanical energy from the natural flow of heat to cold, it uses mechanical energy to drive an unnatural flow of heat from cold to hot.

and cold objects *increases* in this process. The whole purpose of a refrigerator, on the other hand, is to make heat energy flow from cold to hot. How can we do this without violating the second law?

It turns out that we can do this, but we have to *supply* mechanical energy to make this happen. The schematic diagram of a refrigerator presented in figure T9.7 makes it clear that a refrigerator is essentially a heat engine operating in reverse (compare with figure T9.2). Instead of tapping the natural flow of energy from hot to cold to produce a bit of mechanical energy, a refrigerator uses mechanical energy to *drive* the unnatural flow of heat flow from cold to hot!

In the absence of the mechanical energy input, a flow of heat energy from cold to hot certainly *would* violate the second law. If energy $|Q_C| = |Q_H|$ were just to flow directly from the cold reservoir to the hot reservoir, the entropy changes of those reservoirs would be

Second law restrictions on refrigerator performance

$$\Delta S_C = -\frac{|Q_C|}{T_C} \qquad \Delta S_H = +\frac{|Q_H|}{T_H} \qquad \text{(T9.8)}$$

Since $|Q_C| = |Q_H|$ if no mechanical energy is involved and $T_C < T_H$, this implies that $|\Delta S_C| > |\Delta S_H|$, meaning that the *loss* in entropy of the cold reservoir is greater than the *gain* in entropy of the hot reservoir. Thus the total entropy of the interacting reservoirs decreases, in violation of the second law.

But we can prevent violation of the second law by adding enough mechanical energy that $|Q_H|$ becomes enough larger than $|Q_C|$ so that

$$\frac{|Q_H|}{T_H} \geq \frac{|Q_C|}{T_C} \qquad \text{meaning that} \qquad \Delta S_{\text{TOT}} \geq 0 \qquad \text{(T9.9)}$$

By supplying mechanical energy to the hot reservoir, therefore, we are essentially giving it enough extra entropy to make up for the otherwise larger entropy loss of the cold reservoir.

Exercise T9X.9

Multiply both sides of equation T9.9 by T_H and subtract $|Q_C|$ from both sides to show that

$$|Q_H| - |Q_C| \geq \left(\frac{T_H}{T_C} - 1\right)|Q_C| \qquad (T9.10)$$

Note that by conservation of energy, $|Q_H| - |Q_C| = |W| = $ the mechanical energy that we have to supply. This equation thus specifies how large $|W|$ must be.

How can we most usefully quantify a refrigerator's effectiveness? We are most interested in how much heat $|Q_C|$ the refrigerator can extract from the cold reservoir (the benefit) for a given amount of work $|W|$ (the cost). We call the ratio of these quantities the refrigerator's **coefficient of performance (COP)**:

Definition of coefficient of performance (COP)

$$\text{COP} \equiv \frac{\text{benefit}}{\text{cost}} = \frac{|Q_C|}{|W|} \qquad (T9.11)$$

If we combine equations T9.10 and T9.11, we see that

The limit on refrigerator performance

$$\text{COP} \equiv \frac{|Q_C|}{|W|} \leq \frac{T_C}{T_H - T_C} \qquad (T9.12)$$

Purpose: This equation describes the limit that the second law of thermodynamics imposes on the coefficient of performance of a refrigerator that uses work $|W|$ to extract heat $|Q_C|$ from a cold reservoir at temperature T_C and exhausts energy $|Q_H|$ to a hot reservoir at temperature T_H.

Limitations: Both T_H and T_C must be absolute temperatures. The equation assumes that the refrigerator operates between well-defined reservoirs.

Exercise T9X.10

Verify that equation T9.12 is correct.

Example T9.1

Problem Estimate the maximum possible coefficient of performance for a standard kitchen refrigerator.

Model and Translation In the case of an ordinary kitchen refrigerator, the cold reservoir (the inside of the refrigerator) has a temperature of about

$40°F \approx 5°C \approx 278$ K, while the hot reservoir (the kitchen itself) has a temperature equal to room temperature ≈ 295 K, about 17 K warmer.

Solution   The maximum theoretical COP for such a refrigerator would be about $(278 \text{ K})/(17 \text{ K}) = 16.4$.

Evaluation   Note that the COP comes out unitless, which is appropriate for a ratio of energies. This means that such a refrigerator could (in principle) remove an amount of heat from its interior that is more than 16 times larger than the mechanical energy we put in!

Note that COP values will be typically greater than 1! We can take advantage of this fact to construct an effective home heating device.

Example T9.2

Problem   A **heat pump** (see figure T9.8) is essentially a refrigerator that moves heat from outside the house (the cold reservoir) to inside the house (the hot reservoir). Imagine that the outside temperature is 0°C and the inside is 22°C, and that the COP of the heat pump is 8.0. (a) Show that this COP is possible. (b) If the house requires 36 kW of heat energy to keep its temperature at 22°C, how much energy do we have to supply to the heat pump?

(a) *Model and Solution*   Since $T_C = 0°C = 273$ K and $T_H = 22°C = 295$ K, the maximum possible COP for the heat pump is $T_C/(T_H - T_C) = (273 \text{ K})/(22 \text{ K}) = 12.4$. Therefore, a COP of 8.0 is certainly possible.

(b) *Model*   The problem states that we have to supply an energy $|Q_H| = 36,000$ J to the house every second. Since $|Q_H| - |Q_C|$ is equal to the mechanical energy W that we have to supply, and since equation T9.13 implies that $|Q_C| = (COP)W$, we have

$$W = |Q_H| - |Q_C| = |Q_H| - (COP)W \quad \Rightarrow \quad W(1 + COP) = |Q_H| \tag{T9.13}$$

Solution   Solving for W and plugging in the numbers, we get

$$\Rightarrow \quad W = \frac{|Q_H|}{1 + COP} = \frac{36 \text{ kJ}}{9} = 4 \text{ kJ} \tag{T9.14}$$

Evaluation   This is the mechanical energy that we have to supply each second. Therefore, if we use a heat pump, a mere 4 kJ of mechanical energy will bring 36 kJ of heat energy into the house, a real bargain!

If everyone would use heat pumps to heat their houses in the winter, a very large amount of energy could be saved. Unfortunately, heat pumps are quite a bit more expensive (and somewhat less reliable) than standard furnaces (it is still relatively inexpensive to heat a home with natural gas, even if supplying 36 kJ of heat requires 36 kJ worth of natural gas). This makes heat pumps less economically feasible than they should be, considering the value of conserving energy and producing less greenhouse gases.

Figure T9.8
The condenser unit of a heat pump used in home heating.

T9.6   The Carnot Cycle

The **Carnot cycle,** which was first described by the French physicist Sadi Carnot (pronounced *car-NOH*) in 1824, is a hypothetical cyclic process that can be used to convert heat to mechanical energy by using an ideal gas as the working substance. While this particular cycle is not used in realistic engines (for a variety of reasons), it does describe an idealized sequence of gas processes that could be used in principle to produce mechanical energy at the maximum theoretical efficiency allowed by the second law of thermodynamics.

Steps in the Carnot cycle

The Carnot cycle uses an ideal gas confined in a cylinder by a piston, as shown in figure T9.9. A massive flywheel (see figure T9.10a) keeps the piston moving in and out of the cylinder at a regular pace and stores the mechanical

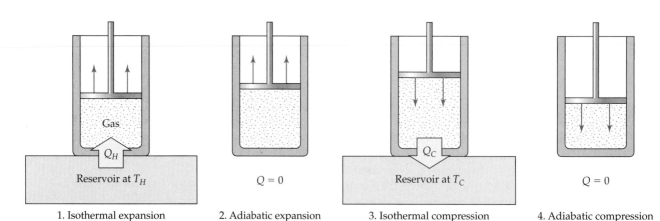

Gas		Q_C	
Q_H	$Q = 0$		$Q = 0$
Reservoir at T_H		Reservoir at T_C	
1. Isothermal expansion	2. Adiabatic expansion	3. Isothermal compression	4. Adiabatic compression

Figure T9.9
The Carnot cycle. You should imagine that the piston is connected to a flywheel (see figure T9.10a) that moves the piston in and out and stores the mechanical energy produced by the cycle.

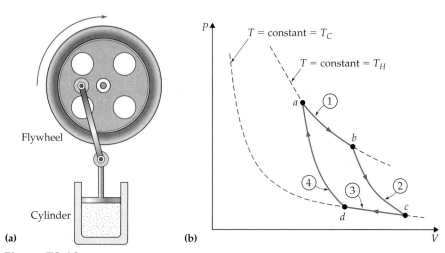

Figure T9.10
(a) This drawing shows how the flywheel is attached to the piston in such a way that it moves the piston in and out of the cylinder as it turns. (b) A P–V diagram of the Carnot cycle process. The net work energy flowing out of the gas in the cycle is the area enclosed by the process on the diagram.

energy produced by the engine. The Carnot cycle consists of the four steps shown in figure T9.9. A detailed description of the steps follows.

1. Just as the flywheel begins to pull the piston out, we put the cylinder in thermal contact with a reservoir at temperature T_H, which keeps the gas temperature fixed at T_H. Since the thermal energy U of an ideal gas depends on N and T but not on V, this means that U also is fixed. However, the expanding gas does work on the piston, so to keep U fixed, heat must flow into the gas from the reservoir. Let this amount of heat be $|Q_H|$.
2. Now we remove the reservoir. The gas expands adiabatically as the piston continues to move out. Work energy continues to flow out of the gas as it expands, but no heat is coming in to replace it, so the gas's thermal energy, and thus its temperature, decreases. The expansion continues until the gas temperature falls to T_C and the flywheel has pulled the piston to its maximum outward position.
3. The flywheel now begins to push the piston *into* the cylinder while the gas is in contact with the cold reservoir at temperature T_C. Because work energy flows *into* the gas as it is compressed, but its temperature (and thus its thermal energy) does *not* increase, an amount of heat energy equal to the work energy flowing in must flow out of the gas into the cold reservoir. Let us call this amount of heat $|Q_C|$.
4. Finally, the reservoir is removed. The piston continues to compress the gas (now adiabatically). Now work energy flows into the gas, but no heat flows out, so the thermal energy and temperature of the gas increase. The compression continues until the gas temperature reaches T_H, and the piston has moved to its innermost position. The gas is now in the same state as when the cycle started.

In a certain sense, the gas is used in this process to carry heat from the hot reservoir to the cold reservoir. In this process, however, there is a net flow of mechanical energy out of the gas: the gas converts *some* of the heat it carries to the cold reservoir to mechanical energy. How do we know? Consider the P–V diagram shown in figure T9.10b. The two curves of constant $T = T_H$ and $T = T_C$ are shown as dotted curves. These are curves where $P \propto 1/V$, since the

Why this cycle produces net mechanical energy

ideal gas law implies that when T is constant, $PV = $ constant. The relationship between P and V during an adiabatic process is given by equation T3.23a:

$$PV^\gamma = \text{constant} \qquad \gamma = 1 + \frac{2}{f} \qquad \text{(T9.15)}$$

where f is the number of degrees of freedom available to each gas molecule (f is 3 for a monatomic gas and 5 for a diatomic gas). This equation implies that $P \propto V^{-\gamma}$, and since $\gamma > 1$, this means that P falls off more sharply with increasing volume than the curves for the isothermal process, as shown.

Remember that the work done on or by a gas in a given quasistatic process is equal to the area under the curve representing that process on a P–V diagram. The diagram clearly shows that the area under the curves of processes 1 and 2 is greater in magnitude than the area under the curves of processes 3 and 4. The mechanical energy moving *out* of the gas during the expansion processes is thus greater than the mechanical energy moving into the gas during the compression processes. Thus, during the cycle as a whole, mechanical energy leaves the gas (and is stored in the flywheel). This must come at the expense of some part of the heat energy that it absorbed in the first step.

The efficiency of a Carnot engine

One of the main reasons that the Carnot cycle is of interest is that such an engine (in the absence of friction or other imperfections) would operate at the maximum efficiency allowed by the second law. One can show directly from formulas we derived in chapter T3 for the work done in isothermal and adiabatic processes that $e = (T_H - T_C)/T_H$ for a Carnot engine (see problem T9S.14). But we can arrive at the same result more quickly by considering the entropies involved. During the adiabatic steps, the reservoirs are disconnected, and we know the gas's entropy does not change during an adiabatic volume change, so no new entropy is created in either of these steps. During either of the isothermal processes, the gas and the reservoir have the same temperature, so a heat flow between them also does not create any new entropy (the entropy in one simply flows to the other). So the net change in entropy of the system consisting of the reservoirs and the engine is *zero* for a complete cycle, implying that the inequalities in equations T9.5, T9.6, and T9.7 apply, implying that the Carnot engine's efficiency is the maximum allowed by the second law.

One can create a Carnot refrigerator simply by running the Carnot cycle in reverse, that is, by compressing the gas during steps 1 and 2 and expanding it in steps 3 and 4. Figure T9.11 shows a P–V diagram and an energy flow

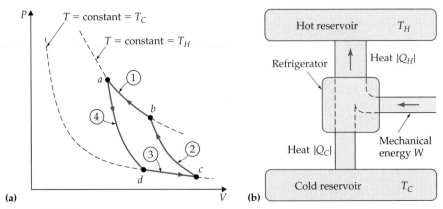

(a)

(b)

Figure T9.11

(a) A P–V diagram for a Carnot engine that we force to run in reverse. (b) The net effect of the cycle is to extract heat energy from the cold reservoir and deposit heat into the hot reservoir.

for the resulting process. Since, again, this generates no new entropy, this refrigerator must have the maximum COP allowed by the second law.

While the Carnot cycle is maximally *efficient*, it is not very practical. To keep from generating new entropy in the isothermal steps, the temperature difference between the gas and the reservoirs must be as small as possible. But the smaller the temperature difference, the slower the heat will flow, so it takes a very long time to get much energy out of such an engine. More realistic engines make compromises between efficiency and power (work per time) generated, which is one reason that realistic engines are less than maximally efficient. Some more realistic engines are discussed in the problems.

TWO-MINUTE PROBLEMS

T9T.1 Classify the following hypothetical perpetual motion machines as being perpetual motion machines of the first kind (A) or the second kind (B).
  a. An electric car runs off a battery which drives the front wheels. The car's rear wheels drive a generator which recharges the battery.
  b. An electric car runs off a battery. When the driver wants to slow down the car, instead of applying the brakes he or she throws a switch that connects the wheels to a generator that converts the car's kinetic energy back to energy in the battery.
  c. An engine's tank is filled with water. When the engine operates, it slowly freezes the water in the tank, converting the energy released to mechanical energy.
  d. Compressed air from a tank blows on a windmill. The windmill is connected to a generator that produces electrical energy. Part of that electrical energy is used to compress more air into the tank.
  e. A normal heat engine is used to drive an electric generator. Part of the power from this generator runs a refrigerator that absorbs the waste heat from the engine and pumps it back into the hot reservoir.

T9T.2 In a maximally efficient heat engine, the amount of entropy that the hot reservoir loses as heat flows out of it is exactly balanced by the entropy that the cold reservoir gains as the waste heat flows into it (T or F).

T9T.3 A heat engine produces 300 W of mechanical power while discarding 1200 W into the environment (its cold reservoir). What is the efficiency of this engine?
  A. 0.20
  B. 0.25
  C. 0.33
  D. Other (specify)

T9T.4 Imagine that in Iceland, scientists discover geothermal vents that produce abundant pressurized steam with a temperature of 300°C. Engineers construct a heat engine that uses this steam as a hot reservoir and a nearby glacier as a cold reservoir. What is the maximum possible efficiency of this engine?

  A. 300%
  B. 100%
  C. 52%
  D. 45%
  E. 22%
  F. Other

T9T.5 Imagine that you are trying to design a personal fan that you wear on your head and operates between your body temperature (37°C) and room temperature (22°C). What is the maximum possible efficiency of this device?
  A. 170%
  B. 95%
  C. 54%
  D. 46%
  E. 5%
  F. Other (specify)

T9T.6 As the temperature difference between its reservoirs increases, the COP of a refrigerator
  A. Gets larger
  B. Gets smaller
  C. Does not change

T9T.7 A refrigerator uses 100 W of electric power and discards 600 W of thermal power into the kitchen. What is its coefficient of performance?
  A. 0.17
  B. 0.20
  C. 5
  D. 6
  E. Other (specify)
  F. Impossible because it violates conservation of energy

T9T.8 Someone comes to your house selling a device that draws heat from groundwater and supplies that heat to your house. The salesperson claims that the amount of heat energy entering the house will far exceed the electrical energy supplied to the device. What the salesperson claims here is physically impossible (T or F).

HOMEWORK PROBLEMS

Basic Skills

T9B.1 We have seen that heat energy cannot be entirely converted to mechanical energy (work). Can mechanical energy be entirely converted to heat? If so, give some examples. If not, explain why not.

T9B.2 A heat engine operates between a hot reservoir at 950°C and a cold reservoir at roughly room temperature (22°C). What is its maximum possible efficiency?

T9B.3 An engine uses water at 100°C and at 0°C as hot and cold reservoirs, respectively. What is its maximum possible efficiency?

T9B.4 A certain heat engine extracts heat energy at a rate of 600 W from a hot reservoir, and it discards energy at a rate of 450 W to its cold reservoir. What is its efficiency?

T9B.5 Why is it important that an air conditioner (which is just a kind of refrigerator) be placed in a window or otherwise have access to the hot environment outside? Would not the air conditioner be more effective if it were placed in the center of the room?

T9B.6 Can you cool your kitchen by leaving the refrigerator door open? Can you heat your kitchen by leaving the oven door open? Explain.

T9B.7 A certain air conditioner maintains the inside of a room at 20°C (68°F) when the temperature outside is 37°C (99°F). What is its maximum possible COP?

T9B.8 A refrigerator with a COP of 8.4 uses 300 W of electric power to extract heat energy from the interior of the refrigerator, which is at 40°F (5°C). At what rate is heat energy removed from its interior?

Synthetic

T9S.1 Imagine that you wish to increase the efficiency of a heat engine. Other things being equal, would it be better to increase the hot reservoir's temperature T_H or decrease the cold reservoir's temperature T_C by the same amount? Explain your response.

T9S.2 A hydroelectric power plant, which converts the gravitational potential energy of water stored behind a dam to electrical energy, can operate at very nearly 100% efficiency. How can it do this when a nuclear power plant cannot operate at better than about 40% efficiency?

T9S.3 Imagine that your local power company claims that "heating your home with *electricity* is 100% efficient." In what sense is this true? In what sense is it misleading?

T9S.4 The following wry versions of the first and second laws of thermodynamics have circulated in the physics community for a long time:

> The first law: You can't win.
> The second law: You can't even break even.

Explain in your own words what these restated laws mean.

T9S.5 An old friend comes to you, asking you to invest in the production of a great new toy: a rubber ball that bounces higher with each succeeding bounce (the kinetic energy released during each bounce comes from the internal energy of the rubber). Explain why you should decline this offer.

T9S.6 The refrigerator COP does *not* express the benefit-to-cost ratio for a heat pump used to heat a building.
(a) What ratio does?
(b) Use an argument similar to that leading from equations T9.8 to T9.12 to determine the limit that the second law imposes on this ratio.

T9S.7 The temperatures generated in the cylinder of an automobile engine can be in excess of 1500 K. Estimate the theoretical maximum possible efficiency of such an engine. (*Hint:* What is T_C going to be, roughly?)

T9S.8 Imagine a power plant designed to exploit the temperature difference between the ocean surface ($\approx 30°C$) and the ocean floor ($\approx 4°C$). Show that a power plant producing 1000 MW of power would have to dump more than 10,000 MW of waste energy into the cold ocean water.

T9S.9 A freezer with a coefficient of performance of 4.9 uses 250 W of power. How long would it take such a freezer to freeze 15 kg of water?

T9S.10 Imagine that a site in Iceland is discovered where temperatures a relatively short distance below the surface of the earth are in the range of 600°C (due to the relatively close proximity of molten rock upwelling from deep inside the earth). A geothermal power plant is constructed at this site. The cold reservoir for this plant is a pool of water constantly fed with ice from a glacier. If the plant produces 100 MW of electrical energy, what is the rate at which ice is melted?

T9S.11 Imagine a solar electric power plant that operates by using a gigantic mirror to concentrate light on the boiler for a steam engine. Assume that the boiler generates steam at a temperature of 550°C. The only possibility for a cold-temperature reservoir is a nearby creek that has an average width of 4 m and an average depth of 0.7 m, flowing at a rate of 0.7 m/s. What is the maximum electric power that could be produced by the plant if it boils the entire creek dry?

T9S.12 Let us use a heat pump to cook dinner! Say that the temperature of your kitchen is 23°C. You set up a heat pump that has 45% of its theoretical maximum COP, and you want to be able to supply 700 W of heat energy for boiling soup. How much electrical energy do you have to supply to the heat pump? How much better is this (if any) than simply using an electric hot plate?

T9S.13 A perfect refrigerator would move a certain amount of heat from a cold reservoir to a hot reservoir without using any mechanical energy (in obvious violation of the second law of thermodynamics). Using energy flow diagrams like those shown in figures T9.1 and T9.10b, argue that if you had a perfect heat engine, you could combine it with a realistic refrigerator to create a perfect refrigerator.

T9S.14 In this problem, you will calculate the efficiency of a Carnot engine *without* referring to entropy.

(a) Consider the isothermal expansion in step 1 of the Carnot cycle. Because the temperature of the gas remains constant, work energy that flows out of the gas as it expands must be balanced by the heat energy flowing into the gas. Use this and equation T3.12 for the work done in an isothermal expansion to show that

$$|Q_H| = +Nk_B T_H \ln\left(\frac{V_b}{V_a}\right) \qquad (T9.16)$$

(b) In a similar manner, show that for step 3 we have

$$|Q_C| = +Nk_B T_C \ln\left(\frac{V_c}{V_d}\right) \qquad (T9.17)$$

[*Hint:* Be very careful with signs. Remember that the absolute values imply that $|Q_H|$ and $|Q_C|$ must be positive, and also remember that $\ln(1/x) = -\ln x$.]

(c) Now consider the adiabatic processes. According to equation T3.23a, $TV^{\gamma-1}$ is constant for an adiabatic process. Argue that this means that

$$T_H V_b^{\gamma-1} = T_C V_c^{\gamma-1} \qquad \text{and} \qquad T_H V_a^{\gamma-1} = T_C V_d^{\gamma-1} \qquad (T9.18)$$

(d) Show that this implies that

$$\frac{V_c}{V_d} = \frac{V_b}{V_a} \qquad (T9.19)$$

(e) Finally, combine equations T9.16, T9.17, and T9.19 to determine the efficiency of the Carnot engine.

T9S.15 Figure T9.12 illustrates the operation of an idealized automobile engine. We can consider each cycle of the engine to be divided into five steps.

1. The piston moves down the cylinder, drawing in a mixture of gasoline and air through a valve that is opened at the top of the cylinder. We call this step the *intake stroke*.
2. The valve is closed, and the piston moves back up the cylinder, compressing the gasoline/air mixture roughly adiabatically. This step is called the *compression stroke*.
3. When the gas is fully compressed, a spark from the spark plug at the top of the cylinder causes the gasoline/air mixture to explode. (In a diesel engine, the increase in temperature caused by the adiabatic compression itself ignites the gas; no spark plug is necessary.)
4. The hot gases produced by the explosion drive the piston down (this is the power stroke). The exhaust gases expand adiabatically during this step.
5. Finally, a valve is opened, and the upward-moving piston pushes the exhaust gases out of the cylinder (this is the exhaust stroke). This returns the engine to its original state.

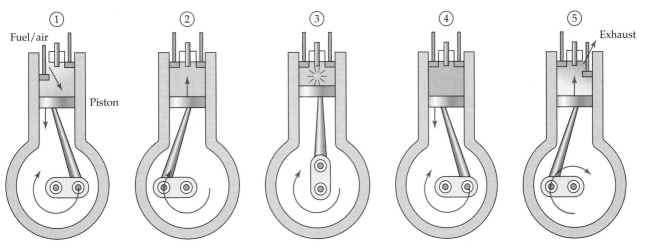

Figure T9.12
The five steps in an automobile engine cycle.

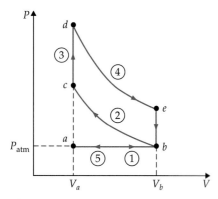

Figure T9.13

An idealized *P–V* diagram for an automobile engine cycle. Step $e \rightarrow b$ happens when the exhaust valve opens freely into the atmosphere. Note also that during steps 1 and 5 the number of molecules in the cylinder is not constant.

Figure T9.13 shows an idealized *P–V* diagram for an automobile engine. During the intake stroke (step 1), the volume of the cylinder changes from V_a to V_b, drawing in the gasoline/air mixture at atmospheric pressure and ambient temperature. At point b, the valve closes and the gas is adiabatically compressed to volume V_a again (point c); this is the compression stroke. The gasoline is then ignited, and the temperature of the gas (and thus its pressure) suddenly increases while the piston is almost at rest. The power stroke (step 4) involves an adiabatic expansion of the exhaust gases back to volume V_b. At point e, the exhaust valve is opened, and the pressure in the cylinder suddenly drops to atmospheric pressure, taking the exhaust gas back to point b again. (The explosive decompression during this step is why gasoline engines are noisy and require mufflers.) Finally, the exhaust gas is pushed out of the cylinder at atmospheric pressure during the exhaust stroke (step 5). Note that the number of molecules N in the cylinder changes during steps 1 and 5 but is basically constant during the other steps.

(a) Note that heat energy enters the system during step 3 (from the burning gasoline) and leaves the system in step 5 (carried away by the exhausted gases). When the cylinder is finally sealed at the end of step 1, the gas at temperature T_e at the end of step 4 has been replaced by an equal amount of gas at temperature T_b, so the net effect of steps 1 and 5 is the same as if we had not let any gas escape but rather cooled it from temperature T_e to temperature T_b. Argue, therefore, that the heat flowing into and out of the engine is

$$|Q_H| = \frac{5}{2} N k_B (T_d - T_c)$$

and $\quad |Q_C| = \frac{5}{2} N k_B (T_e - T_b) \qquad$ (T9.20)

(b) Use this to show that the engine's efficiency is

$$e = 1 - \frac{T_e - T_b}{T_d - T_c} \qquad (T9.21)$$

(c) Use the fact that $T V^{\gamma-1}$ during an adiabatic process to show that

$$\frac{T_e - T_b}{T_d - T_c} = \left(\frac{V_a}{V_b} \right)^{2/5} \qquad (T9.22)$$

(d) Find an automobile's efficiency if its compression ratio $V_b / V_a = 8$.

Rich-Context

T9R.1 Is it more economical to use a heat pump to heat your home during the winter in California or in Minnesota? Assume that energy costs and house designs are the same in both locations. Also assume that the rate at which a house loses energy is proportional to the temperature difference between the outside and inside of the house.

T9R.2 The *Stirling engine* is a practical heat engine (invented in 1816 by Robert Stirling) that was widely used in the 19th century before being displaced by the internal combustion engine. It is ingeniously simple, involving few moving parts. Although internal combustion engines generate much more power for a given working volume, Stirling engines are still used in applications where their robust reliability and ability to use any external heat source are valuable.

Figure T9.14 illustrates the Stirling cycle and figure T9.15 shows an actual toy Stirling engine. The piston and the displacer are both connected to a flywheel in such a way that when the piston is at rest, the displacer is moving, and vice versa. In the first step, air at the hot end of the heat exchanger absorbs heat from the external heat source and expands, pushing around the displacer and against the piston, driving it out. In step 2, the piston comes essentially to rest, and the displacer moves toward the hot end of the exchanger, displacing the air to the cold end of the exchanger. In step 3, the flywheel begins to push the piston in, but now most of the air inside the engine is at the cool end of the heat exchanger, and so it takes less work to compress it than we got out during step 1. Finally, the piston comes to rest again, and the displacer moves toward the cold end, displacing the air back to the hot end, and we are back where we started.

(a) For the sake of simplicity, assume that steps 2 and 4 occur at exactly constant volumes and that all the air has plenty of time to cool to the temperature T_C of the cool end as the displacer moves during step 2 and to warm to the temperature T_H of the hot end as the displacer moves during step 4. Ignore the small amount of air on the wrong side of the displacer in all steps. Draw a *P–V* diagram for this idealized Stirling cycle.

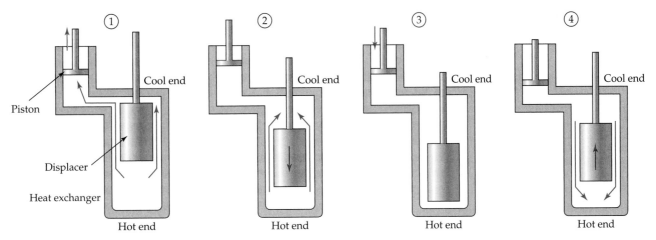

Figure T9.14
A schematic diagram of a Stirling engine.

Figure T9.15
A toy Stirling engine. The fins help the cold end of the heat exchanger keep cool.

(b) Argue, using equation T3.12, that the net work done during this cycle is

$$|W| = Nk_B(T_H - T_C)\ln\left(\frac{V_{max}}{V_{min}}\right) \qquad (T9.23)$$

where V_{max} and V_{min} are the maximum and minimum volumes, respectively, of the gas during the cycle.

(c) Heat flows into the gas in both steps 1 and 4. What is the total heat $|Q_H|$ flowing into the engine during a cycle, expressed in terms of the same quantities? (Remember that air is diatomic.)

(d) Show that the efficiency e of the Stirling engine can be written as

$$\frac{1}{e} = \frac{1}{e_C} + f\left(\frac{V_{max}}{V_{min}}\right) \qquad (T9.24)$$

where $e_C = (T_H - T_C)/T_H$ is the efficiency of a Carnot engine and $f(V_{max}/V_{min})$ is some function of the quantity V_{max}/V_{min} that you should determine. Argue that this means that the efficiency of even our idealized Stirling engine is less than that of a Carnot engine.

(e) What is the efficiency of our idealized Stirling engine when operating between temperatures of 600 K and 300 K, if its compression ratio $V_{max}/V_{min} \approx 3$? Compare this to the efficiency of a Carnot engine.

Advanced

T9A.1 Imagine that you are trying to construct a practical Carnot engine that will deliver 1 horsepower (hp) ≈ 700 W. The parameters are as follows. For safety reasons, your high-temperature reservoir can be no higher than 535°F (280°C). Your engine is cooled by the surrounding air. Because volume changes have to be reasonably quasistatic, you are limited to 2 cycles per second. The metal used for your cylinder can withstand 25 atm of pressure, and various considerations limit the ratio of your maximum-to-minimum volume to be no more than 10. Find the minimum

and maximum volumes for your cylinder. (For comparison, a gasoline engine with a cylinder volume of a few thousand cubic centimeters can produce roughly 100 hp.) (*Hints:* Study problem T9.14. Note that $|W| = e|Q_H|$ and $|Q_H|$ is given by equation T9.23. This will get you started.)

ANSWERS TO EXERCISES

T9X.1 Substituting $W = |Q_H| - |Q_C|$ into equation T9.1 yields

$$e = \frac{|Q_H| - |Q_C|}{|Q_H|} = 1 - \frac{|Q_C|}{|Q_H|} \qquad (T9.25)$$

T9X.2 The second law of thermodynamics requires that the total entropy of the interacting objects here (the two reservoirs and the engine) not decrease:

$$\Delta S_{TOT} = \Delta S_C + \Delta S_{engine} + \Delta S_H \geq 0 \qquad (T9.26)$$

As discussed, $\Delta S_{engine} = 0$ (at least in the long run). Substituting the values of ΔS_H and ΔS_C from equation T9.4 yields the desired result.

T9X.3 Multiplying both sides by $T_C|Q_H|$, we get

$$\frac{|Q_C|}{|Q_H|} - \frac{T_C}{T_H} \geq 0 \qquad (T9.27)$$

Adding T_C/T_H to both sides gives the desired result.

T9X.4 Equation T9.3 reads

$$e = 1 - \frac{|Q_C|}{|Q_H|} \qquad (T9.28)$$

This will be *smaller* than the quantity $1 - T_C/T_H$, because according to equation T9.6, $|Q_C|/|Q_H| \geq T_C/T_H$, so subtracting $|Q_C|/|Q_H|$ from 1 yields a smaller number than subtracting T_C/T_H from 1. Therefore

$$e \leq 1 - \frac{T_C}{T_H} \qquad (T9.29)$$

as stated.

T9X.5 $e \leq (T_H - T_C)/T_H = (573 \text{ K} - 313 \text{ K})/(573 \text{ K}) = 0.45$.

T9X.6 Every second, the plant produces $|W| = 1000$ MJ of mechanical energy. If the efficiency is 0.34, this means that $|Q_H| = |W|/e = (1000 \text{ MJ})/0.34 = 2900$ MJ. This means that $|Q_C| = |Q_H| - |W| = 1900$ MJ. This is the energy budget every second, so the rate of waste heat flow is 1900 W.

T9X.7 The flow rate of the river is $(67 \text{ m})(3 \text{ m})(0.5 \text{ m/s}) = 100 \text{ m}^3/\text{s} = 10^5$ kg/s. In 1 s, then, the 1900 MJ of waste heat produced by the plant has to be distributed in 10^5 kg of water. The resulting increase in temperature is

$$dT = \frac{dU}{mc} = \frac{1.9 \times 10^9 \text{ J}}{(10^5 \text{ kg})(4186 \text{ J kg}^{-1} \text{ K}^{-1})} = 4.5 \text{ K} \qquad (T9.30)$$

T9X.8 The energy required to vaporize a mass m of water is $dU = mL$, where L is the latent heat of vaporization. The amount of water vaporized by 1900 MJ of energy is thus

$$m = \frac{dU}{L} = \frac{1.9 \times 10^9 \text{ J}}{2.257 \times 10^6 \text{ J/kg}} = 840 \text{ kg} \qquad (T9.31)$$

This is less than 1 m^3 of water, so less than 1% of the river would have to be vaporized.

T9X.9 Following the directions, we get

$$\frac{|Q_H|}{T_H} \geq \frac{|Q_C|}{T_C} \quad \Rightarrow \quad |Q_H| \geq \frac{T_H}{T_C}|Q_C|$$

$$\Rightarrow \quad |Q_H| - |Q_C| \geq \frac{T_H}{T_C}|Q_C| - |Q_C|$$

$$= |Q_C|\left(\frac{T_H}{T_C} - 1\right) \qquad (T9.32)$$

T9X.10 Since $W = |Q_H| - |Q_C|$, equation T9.11 implies that

$$\text{COP} = \frac{|Q_C|}{W} = \frac{|Q_C|}{|Q_H| - |Q_C|} \qquad (T9.33)$$

Plugging equation T9.10 into this (and noting that when we *divide* by a bigger number the result is *smaller*), we get

$$\text{COP} \leq \frac{|Q_C|}{(T_H/T_C - 1)|Q_C|} = \left(\frac{T_H}{T_C} - 1\right)^{-1} \qquad (T9.34)$$

If we replace 1 by T_C/T_C, add the terms, and invert the ratio, we get equation T9.12.

Glossary

absolute zero: the temperature at which the pressure in a constant-volume gas thermometer would be exactly zero. This is the lowest possible temperature. (Section T1.5.)

adiabatic index: the quantity γ appearing in the equations $TV^{\gamma-1} = $ constant and $PV^{\gamma} = $ constant that describe adiabatic processes (see below). If f is the number of degrees of freedom the gas molecules have, then $\gamma = 1 + 2/f$. (Section T3.6.)

adiabatic process: a constrained gas process in which heat is not allowed to flow into or out of the gas. (Section T3.4.)

Boltzmann factor: the quantity e^{-E/k_BT}. The probability that a quantum system in thermal contact with a reservoir will be in a quantum state with energy E is proportional to the Boltzmann factor. (Section T6.4.)

Boltzmann's constant k_B: the physical constant $k_B = 1.38 \times 10^{-23}$ J/K $= 8.63 \times 10^{-5}$ eV/K appearing in the ideal gas law. This constant more fundamentally links a molecule's average kinetic energy to the ambient temperature: $K_{\text{avg}} = (3/2)k_BT$. (Section T2.1.)

brownian motion: the jiggling of small particles suspended in a solution due to the impact of water molecules. (Section T2.3.)

Carnot cycle: an idealized sequence of ideal gas processes that produces mechanical energy from heat at the maximum efficiency allowed by the second law. (Section T9.6.)

characteristic temperature T_ε: the quantity ε/k_B, where ε is the characteristic energy difference between the lowest energy levels of a quantum system. This quantity specifies the approximate temperature at which the quantum system begins to store its expected amount of thermal energy (i.e. the temperature at which it "switches on"). (Section T7.4.)

coefficient of performance (COP) (of a refrigerator): defined to be $|Q_C|/|W|$; that is, the amount of heat the refrigerator removes from the cold reservoir (the benefit a refrigerator produces), expressed as a multiple of the mechanical energy we have to give to the refrigerator (the cost). COPs are typically greater than 1. (Section T9.5.)

constant-volume gas thermoscope: a thermoscope that quantifies the temperature of its surroundings in terms of the pressure of a quantity of gas held at a fixed volume. (Section T1.5.)

constrained process: a process that takes a gas through a sequence of states determined by some constraint placed on the gas (such as "P must remain constant"). (Section T3.4.)

diatomic (molecule): a molecule containing two atoms. (Section T2.4.)

efficiency e (of a heat engine): defined to be $W/|Q_H|$; that is, the fraction that the mechanical energy W produced by the engine represents of the total heat energy $|Q_H|$ supplied to the engine. (Section T9.3.)

Einstein solid: a microscopic model for a crystalline solid that treats each atom as if it were held in its lattice position by springs (i.e., the atoms behave as independent identical three-dimensional harmonic oscillators). (Section T4.2.)

entropy S (of a macrostate): defined to be $S = k_B \ln \Omega$, where Ω is the multiplicity of the macrostate. The entropy of a two-object system in a given macropartition is the sum of the entropies corresponding to each object's macrostate in that macropartition: $S_{AB} = S_A + S_B$. (Section T5.4.)

first law of thermodynamics: $\Delta U = Q + W$, which follows from the definitions of heat and work and the law of conservation of energy. (Section T3.1.)

fluctuation: a random change in a combined system's macropartition due to random energy transfers while the system is near equilibrium. (Such fluctuations are possible because macropartitions *very* near the most probable one have nearly the same number of microstates, and thus are about equally probable.) (Section T5.2.)

fundamental assumption (of statistical mechanics): each accessible microstate of a system is equally probable. *Accessible* here means that the microstate must be consistent with the known macrostate of the system. For a combined system of two objects in thermal contact, this means that the probability that a given macropartition will be observed is proportional to its multiplicity. (Section T4.5.)

gas constant R: the product of Boltzmann's constant and Avogadro's number: $R = k_B N_A = 8.31$ J K^{-1} mol^{-1}. (Section T2.1.)

heat Q: energy that flows across a system boundary as the result of a temperature difference *alone*. (Section T3.1.)

heat engine: a device that converts heat energy to mechanical energy, often using a cyclic gas process. Real heat engines tap into the heat energy that naturally flows from a hot reservoir to a cold reservoir (driven by the second law) and convert part of this flow to mechanical energy. (Section T9.1.)

heat pump: a refrigerator that uses mechanical energy to cool the outdoors while warming a house with its "waste" heat. Since the heat moved by a heat pump can be much larger than the mechanical energy supplied, this is an efficient way to warm a house. (Section T9.5.)

ideal gas: a gas that is well described by the ideal gas law (see below). This is an idealization, but gases at normal temperatures and pressures are roughly ideal, and the approximation improves as the gas density decreases. (Section T2.1.)

ideal gas law: an empirical law $PV = Nk_BT$ that links the pressure P, volume V, number of molecules N, and temperature T of a gas. This law is an idealization that all real gases approach in the limit of zero density. (Section T2.1.)

irreversible process: a process involving an isolated system that is physically impossible if the system's macroscopic initial and final conditions are reversed. A movie of such a process will look very strange if reversed. (Section T1.2.)

isobaric process: a constrained gas process in which the gas's pressure P is held constant. (Section T3.4.)

isochoric process: a constrained gas process in which the gas's volume V is held constant. (Section T3.4.)

isothermal process: a constrained gas process in which the gas's temperature is held constant. (Section T3.4.)

kelvin: the SI unit of temperature. The definition of the Kelvin scale makes 1 K equal in size to 1°C of temperature difference. (Section T1.5.)

Kelvin temperature scale: the internationally accepted temperature scale for physics, defined to be proportional to the pressure of a constant-volume gas thermoscope (in the limit of zero gas density). The constant of proportionality is fixed by defining the temperature of the triple point of water to be 273.16 K. (This makes the freezing point of water at standard pressure 273.15 K.) (Section T1.5.)

latent heat L: a substance's change in thermal energy per unit of mass during a phase change: $\Delta U = \pm mL$. ΔU is positive if the substance goes from solid to liquid, liquid to gas, or solid to gas; and negative if the transformation goes in the other direction. (Section T8.4.)

law of Dulong and Petit: the heat capacity of a monatomic crystalline solid is $dU/dT \approx 3Nk_B$. (Section T2.5.)

macropartition: a *pair* of macrostates for two objects in thermal contact consistent with the overall macrostate for the combined system. In the case of a pair of Einstein solids,

specifying the macropartition amounts to specifying the energies U_A and U_B of each solid, subject to the constraint that $U = U_A + U_B$ be a fixed constant. (Section T4.4.)

macropartition table: a table listing the possible macropartitions of a system involving two interacting objects and their multiplicities. Table T4.1 shows an example macropartition table. (Section T4.4.)

macroscopic properties: the properties of a complex object that can be measured without knowing anything about the quantum states of its molecules. (Section T3.3.)

macrostate: the state of a system as described by its macroscopic properties. Two systems have the same macrostate if all macroscopically measurable quantities (such as P, V, N, T, U, and so on) have the same values. For an ideal gas, specifying P, V, N, and the type of gas is sufficient to specify the macrostate; the gas's other macroscopic properties are determined by these. (Section T4.1.)

Maxwell-Boltzmann distribution function $\mathscr{D}(v)$**:** the function that when multiplied by dv/v_P yields the probability that a molecule will have a speed in a range of width dv centered on the speed v. Note that $v_P \equiv (2k_BT/m)^{1/2}$ is a constant with units of speed. (Section T7.1.)

microstate: the state of a system specified by describing the quantum state of each molecule in the system. (Section T4.1.)

monatomic (gas)**:** a gas whose molecules consist of a single atom each. (Section T2.4.)

molar mass M_A: the mass (in grams) of Avogadro's number (1 mol) of molecules of a given substance. This mass is *roughly* equal to the number of protons and neutrons in the molecule's atomic nuclei. (Section T2.1.)

multiplicity Ω: the number of microstates consistent with a given macrostate. (Section T4.3.)

n factorial ($n!$)**:** a shorthand expression for the product $n \cdot (n-1) \cdot (n-2) \cdots 2 \cdot 1$. (Section T4.3.)

paradigmatic thermal process: the process of an initially hot object coming into thermal equilibrium with an initially cold object. This simple process raises most of the fundamental questions about temperature, energy, and irreversibility that we seek to answer in this unit. (Section T1.3.)

partial pressure: the part of the total pressure exerted by a gas mixture contributed by one of the gases. (Section T2.1.)

pascal: the SI unit of pressure. $1\,\mathrm{Pa} \equiv 1\,\mathrm{N/m^2}$. (Section T1.5.)

perpetual motion machine: a hypothetical device that would produce endless amounts of mechanical energy. A "perpetual motion machine of the first kind" violates energy conservation by creating energy from nothing. A "perpetual motion machine of the second kind" violates the second law of thermodynamics by extracting energy from a reservoir in such a way as to make the total entropy of the engine and reservoir decrease. (Section T9.1.)

phase change: the transformation of a substance from one *phase* to another. Substance phases most commonly seen are *solid*, *liquid*, and *gas*. (Section T8.4.)

polyatomic (molecule): a molecule constructed of more than two atoms. (Section T2.4.)

pressure (of a fluid) P: the force per unit area exerted by the fluid on a surface separating the fluid from a vacuum. (Section T1.5.)

P–V diagram: a plot of the gas's pressure versus its volume. If N is fixed and known, the macroscopic state of an ideal gas in equilibrium at a given time is represented by a *point* on such a diagram. (Section T3.4.)

quasistatic gas process: a gas process that occurs so slowly that the gas remains uniform and essentially in equilibrium all the time. As the state of the gas changes during a quasistatic process, it traces a curve on a P–V diagram. (Section T3.4.)

replacement process: a hypothetical quasistatic process that we use to compute the entropy change of a system during a process involving non-quasistatic work. Since a system's entropy depends only on its macrostate, if we can think of a quasistatic replacement process that takes the system between the same initial and final macrostates as the non-quasistatic process, the entropy change for both processes should be the same. (Section T8.6.)

reservoir: an object so massive that it can absorb or provide substantial amounts of heat without a discernible change in its temperature. (Section T6.4.)

reversible process: a process involving an isolated system that remains physically possible if the system's macroscopic initial conditions and final conditions are reversed. A movie of such a process will look reasonable if shown in reverse. (Section T1.2.)

root-mean-square (rms) speed v_{rms}: the square root of the average squared speed: $v_{rms} \equiv ([v^2]_{avg})^{1/2}$. (Section T2.3.)

second law of thermodynamics: the entropy of an isolated system never decreases. In the language of chapter T5, this is so because the probability of a large system evolving by chance from one macropartition to another with a significantly smaller multiplicity (and thus entropy) is so small as to be zero for all practical purposes. (Section T5.5.)

specific heat c: the quantity c in the relationship $dU = mc\,dT$ that characterizes the relationship between changes in

an object's temperature and changes in its thermal energy. "Heat" is a horrible misnomer here: we are really talking about *thermal energy*. The quantity c is independent of the object's mass, depends weakly on the object's temperature, but strongly depends on the substance that the object is made of and the phase of that substance. (Section T1.6.)

standard air pressure: 101.3 kPa, approximately equal to the average pressure exerted by the earth's atmosphere at sea level. (Section T1.5.)

statistical mechanics: the theory that explains the laws of thermodynamics in terms of the statistical behavior of the microscopic particles in the objects involved. (Section T1.1.)

temperature T: a physical property of a macroscopic object that can be measured by a thermoscope (see below) and is linked both to the object's thermal energy and to thermal equilibrium. (Section T1.4.)

thermal energy U: the part of a substance's internal energy that can be changed by thermal processes. (Section T3.1.)

thermal equilibrium: the final state of two objects placed in contact. As the objects approach equilibrium, macroscopic properties linked with internal energy and temperature may change with time; but when equilibrium is attained, these properties no longer change: the objects' macroscopic characteristics become stable. (Section T1.3.)

thermal physics: the general category that embraces both thermodynamics and statistical mechanics. (Section T1.1.)

thermodynamics: the study of how temperature, thermal energy, entropy, and other *macroscopic* characteristics of an object are affected by interactions with its surroundings. Thermodynamics describes the general laws governing such interactions without offering any particular explanation for these laws. (Section T1.1.)

thermometer: any thermoscope calibrated to read the same temperature that a constant-volume gas thermoscope would under the same conditions. (Section T1.5.)

thermoscope: a device that quantifies temperature in terms of changes in some measurable physical property of the working substance in the thermoscope. A bar of metal whose resistance changes with temperature is an example of a thermoscope. (Section T1.4.)

triple point of water: a specific temperature and pressure where water can coexist in the solid, liquid, and gas phases simultaneously. This triple point represents a precisely defined physical reference point for defining a temperature scale. (Section T1.5.)

work W: any energy flowing across a system boundary that is *not* heat. (Section T3.1.)

zero-point energy: the energy that a quantum system has when it is in its lowest possible quantum energy state. An Einstein solid has a zero-point energy of $3N\hbar\omega/2$, which, since it cannot be changed and does not depend on temperature, is irrelevant to its thermal behavior. (Section T4.2.)

zeroth law of thermodynamics: the law that asserts that two objects will be in equilibrium *if and only if* they have the same temperature. (Historically, this law was first stated long *after* the laws known as the first, second, and third laws of thermodynamics; but as it is *logically* prior to these laws, it was named the *zeroth* law instead of the fourth law of thermodynamics.) (Section T1.4.)

Index

Periodic Table of the Elements

Key:

1	Atomic number
H	Symbol
1.008	Atomic mass

1 1A	2 2A	3 3B	4 4B	5 5B	6 6B	7 7B	8 8B	9 8B	10	11 1B	12 2B	13 3A	14 4A	15 5A	16 6A	17 7A	18 8A
1 **H** 1.008																	2 **He** 4.003
3 **Li** 6.941	4 **Be** 9.012											5 **B** 10.81	6 **C** 12.01	7 **N** 14.01	8 **O** 16.00	9 **F** 19.00	10 **Ne** 20.18
11 **Na** 22.99	12 **Mg** 24.31											13 **Al** 26.98	14 **Si** 28.09	15 **P** 30.97	16 **S** 32.07	17 **Cl** 35.45	18 **Ar** 39.95
19 **K** 39.10	20 **Ca** 40.08	21 **Sc** 44.96	22 **Ti** 47.88	23 **V** 50.94	24 **Cr** 52.00	25 **Mn** 54.94	26 **Fe** 55.85	27 **Co** 58.93	28 **Ni** 58.69	29 **Cu** 63.55	30 **Zn** 65.39	31 **Ga** 69.72	32 **Ge** 72.59	33 **As** 74.92	34 **Se** 78.96	35 **Br** 79.90	36 **Kr** 83.80
37 **Rb** 85.47	38 **Sr** 87.62	39 **Y** 88.91	40 **Zr** 91.22	41 **Nb** 92.91	42 **Mo** 95.94	43 **Tc** (98)	44 **Ru** 101.1	45 **Rh** 102.9	46 **Pd** 106.4	47 **Ag** 107.9	48 **Cd** 112.4	49 **In** 114.8	50 **Sn** 118.7	51 **Sb** 121.8	52 **Te** 127.6	53 **I** 126.9	54 **Xe** 131.3
55 **Cs** 132.9	56 **Ba** 137.3	57 **La** 138.9	72 **Hf** 178.5	73 **Ta** 180.9	74 **W** 183.9	75 **Re** 186.2	76 **Os** 190.2	77 **Ir** 192.2	78 **Pt** 195.1	79 **Au** 197.0	80 **Hg** 200.6	81 **Tl** 204.4	82 **Pb** 207.2	83 **Bi** 209.0	84 **Po** (210)	85 **At** (210)	86 **Rn** (222)
87 **Fr** (223)	88 **Ra** (226)	89 **Ac** (227)	104 **Rf** (257)	105 **Db** (260)	106 **Sg** (263)	107 **Bh** (262)	108 **Hs** (265)	109 **Mt** (266)	110	111	112	(113)	114	(115)	116	(117)	

58 **Ce** 140.1	59 **Pr** 140.9	60 **Nd** 144.2	61 **Pm** (147)	62 **Sm** 150.4	63 **Eu** 152.0	64 **Gd** 157.3	65 **Tb** 158.9	66 **Dy** 162.5	67 **Ho** 164.9	68 **Er** 167.3	69 **Tm** 168.9	70 **Yb** 173.0	71 **Lu** 175.0
90 **Th** 232.0	91 **Pa** (231)	92 **U** 238.0	93 **Np** (237)	94 **Pu** (242)	95 **Am** (243)	96 **Cm** (247)	97 **Bk** (247)	98 **Cf** (249)	99 **Es** (254)	100 **Fm** (253)	101 **Md** (256)	102 **No** (254)	103 **Lr** (257)